BREATH
TAKING

The Power, Fragility, and Future
of Our Extraordinary Lungs

BREATH
TAKING

MICHAEL J. STEPHEN, MD

Atlantic Monthly Press
New York

FIRST EDITION

Published simultaneously in Canada
Printed in Canada

This book is set in 11-pt. Janson LT by Alpha Design & Composition of Pittsfield, NH.

First Grove Atlantic hardcover edition: January 2021

Library of Congress Cataloging-in-Publication data is available for this title.

ISBN 978-0-8021-4931-2
eISBN 978-0-8021-4933-6

Atlantic Monthly Press
an imprint of Grove Atlantic
154 West 14th Street
New York, NY 10011

Distributed by Publishers Group West

groveatlantic.com

21 22 23 24   10 9 8 7 6 5 4 3 2 1

Life and respiration are complementary. There is nothing living which does not breathe nor anything breathing which does not live.

—William Harvey, 1653

CONTENTS

PART III
THE FUTURE: THE LUNGS PROVIDE
A VISION OF WHAT'S TO COME

PART IV
LIFE, LOVE, AND THE LUNGS

Prologue: Lungs = Life

The lungs are a mysterious and even mystical organ. They are our connection to the atmosphere, the organ that extracts the life force we need to exist. We have acknowledged this power for centuries. The Hebrew word *ruach* literally means "breath," but also means "spirit of life." In the book of Job, the prophet's friend Elihu declares, "the Spirit of God has made me, and the breath of the Almighty gives me life."[1] This same concept is embodied in the New Testament, where the apostle John says that Jesus "breathed" on his disciples, giving them the Holy Spirit.[2]

The life-giving power of the breath is acknowledged very early in the Bible: in the second chapter of Genesis, line 7 reads, "the Lord God formed man from the dust of the ground, and breathed into his nostrils the breath of life, and the man became a living being."[3] Ancient Egyptian cultures also recognized the importance of the breath, the evidence of which we see today in the many ancient statues that had their noses broken off but otherwise were left untouched. This defacement was no accident, but a deliberate act by conquering groups to take the life, in this case the breath of life, away from these icons.[4]

Ancient knowledge about the power of the lungs was not limited to the Western world. Buddhism and Hinduism were based on an understanding of the potency of the breath. According to these disciplines, studying and harnessing the breath was the only recognized way to nirvana. Thich Nhat Hanh, a Vietnamese Buddhist monk, summarizes this ancient philosophy well in his 1975 book *The Miracle of Mindfulness*: "Breath is the bridge which connects life to consciousness, which unites your body to your mind."[5]

The emphasis on the breath is not a thing of the past for Eastern religions; breathing continues to occupy a central role in their teachings today. The Hindu name for the breath is *prana*, which, like the western *ruach* and Holy Spirit, is not just a word for air, but an acknowledgment of the breath as the ultimate life force. This knowledge is spreading back to the West through disciplines such as yoga and mindfulness, but also through techniques aimed at improving endurance, and even intimacy. These practices demonstrate that the mind and the heart follow the lungs, not the other way around.

Civilizations throughout history have equated the breath with the soul, using one term to refer to both. In ancient Egypt, it was *ka*; in Zulu, *umoya*; in ancient Greece, *pneuma*; in Hinduism, *prana*. In his 1653 *Lectures on the Whole of Anatomy*, William Harvey, the famous seventeenth-century British physiologist, stated simply but profoundly, "Life and respiration are complementary. There is nothing living which does not breathe nor anything breathing which does not live."[6]

As an organ, the lungs must do an incredible amount of work all day, every day. With an average respiratory rate of fourteen breaths per minute, and an average volume of 500 milliliters of air per breath, a typical adult inhales and exhales 420 liters of air every hour. In one day, the total is roughly 10,080 liters, a tremendous amount of gas for every human on this planet to utilize. Yet, in the absence of lung disease this work is effortlessly integrated into our lives, happening without any conscious effort.

With a signal from the brain, the diaphragm contracts downward, expanding the lungs in an instant. In this way, the breath of life is drawn into the body, and contained in it are millions of oxygen molecules. The lungs seamlessly pass the oxygen off to the red blood cells, which, with the help of the heart, deliver these molecules of life to the cells of the brain, muscle, kidneys, and other organs. Continuing the circuit, carbon dioxide, produced as oxygen is consumed by our tissues, is whisked through our veins and back to the lungs, and then expelled into the atmosphere as the diaphragm now relaxes. It is a beautiful circle of reuse and recycle, appropriately termed circulation, with the lungs as the centerpiece, the lynchpin connecting the body and the outside world.

That oxygen, life, and lungs all came into our world in relatively close succession is no coincidence. Only with oxygen and some means of extracting it are all things possible—thinking, moving, eating, speaking, and loving. Life and the breath are synonymous. Significantly, our arrival into the world as infants is considered a success when we take our first breath outside of our mother's womb, and our leave-taking occurs when our last breath is expelled.

We are not the only ones breathing, of course. Breathing is the mechanism for harnessing the life force all over the Earth, and every organism above the level of microscopic anaerobe respires, including every single fish and animal, as well as all plants. Plants are known as oxygen producers because of photosynthesis, but they, too, constantly breathe, consuming oxygen to fuel their energy needs at the same time photosynthesis is occurring. We all utilize this communal resource called the atmosphere together.

There is something transcendent in the very structure of our respiratory system. It begins in the trachea, the single wide tube that accepts air after it comes in through the mouth or nose. The trachea branches into the bronchi of the right and left lungs, with the airways continuing to divide into increasingly smaller tubes, finally opening, deep in our lungs, into grapelike clusters called alveoli, the actual site of gas exchange. This structure, taken as a whole, resembles a tree, with its trunk and progressively smaller branches emerging into their site of gas exchange, leaves. Other examples of this configuration in nature abound—streaks of lightning converging into a single bolt only to diverge again as they approach the ground; the tributaries of a riverbed unifying into one main waterway; the human body itself, branching from its trunk to arms and legs, then fingers and toes. The lungs tap into something universal in their structure, maximizing uptake of the life force that surrounds all of us.

Science is beginning to investigate in a serious manner something humanity has known for centuries—that the breath can be used to heal the body. Every year more papers are being published on the healing power of the breath. Evidence of improvement in patients with asthma, chronic obstructive pulmonary disease, chronic pain, depression, and even cancer has been demonstrated. The scientific evidence has started to go even deeper, to the level of our blood, and even our genes. In those who practice breathing exercises, levels of inflammatory proteins in the blood are significantly lower, especially under certain types of stress. Mobilizing the power of the breath has also been shown to turn on anti-inflammatory genes and turn off pro-inflammatory ones, including genes that regulate energy metabolism, insulin secretion, and even the part of our DNA that controls longevity.[7] Looking down the generations, those of us who practice breathing exercises today may well pass on more disease-resistant genes to our descendants tomorrow.

There is also something about the lungs—beyond their role in disease prevention—that is essential to our existence and future survival. The lungs are our youngest structure evolutionarily, having developed as our ancestors emerged from the ocean some four hundred million years ago, well after the heart and other organs had evolved. In addition to being the central organ at both birth and death, they are also the body part we must accommodate in the future if we are going to colonize other planets, or even stay alive on this one, with its radically changing climate and the constant threat of respiratory pathogens. Like our other organs, the lungs are under unconscious control of our brain, but unlike our other organs, we can control them consciously if we choose.

With this element of control, the lungs are hailed at present as an important focal point for the health and progress of society as a whole. We live in a time of extraordinary change, with technology and medicine bringing unheard-of advances in the last hundred years. Average human lifespan has recently doubled, the number of people on Earth tripled. At the same time, we have naturally held onto the same emotions of anxiety and distrust that helped us survive at a time when we faced completely different threats than we do now. For us to move forward as a people and as a planet, we are going to have to become more trusting, more cooperative. The lungs are the organ that could help us make this transition.

Though the lungs are a powerful organ, they are nonetheless badly overlooked and ever more threatened today. The heart has stolen the spotlight in our songs and literature as the embodiment of our emotions and passions. The brain is revered as the seat of our thoughts and desires, credited with our success as a species and celebrated for its complexity. The skin is pampered, a reflection of our beauty in youth and wisdom in old age. Our reproductive system offers the magnetism of sex and the miracle of birth. Usually only someone who is short of breath gives the lungs a second thought.

In the medical world, the statistics demonstrating that the lungs are an ignored organ are stark. Every year, lung cancer kills more people than breast, pancreatic, and colon cancer (the next three leading causes of cancer death) *combined*, and yet lung cancer receives about half the funding that breast cancer does from the National Institutes of Health (NIH) and other government agencies.[8,9]

Meanwhile, the outcomes of many lung diseases are devastating. Idiopathic pulmonary fibrosis (IPF) is a scarring disease of the lungs—one that most people have never heard of—that affects thirty thousand patients every year, about the same as cervical cancer. Funding for research on IPF is poor, and to date not a single drug has definitively been shown to extend the lives of people with the disease. The 50 percent survival rate is horrible, with most patients dead within about four years of being diagnosed.[10] This is worse than most cancers. Except, of course, for lung cancer, that other underfunded, under-recognized condition.

The list of other, similarly neglected diseases includes chronic obstructive pulmonary disease (COPD), inhalation injuries, and asthma. A stigma is attached to many lung diseases and is kept there by some strong biases. The most obvious one is the association with smoking, a major causative factor for both lung cancer and COPD. We have demonized not just tobacco, but smokers as well. A subtle but corrosive prejudice also exists against people suffering from asthma, a condition that is falsely linked to inner cities and an unclean lifestyle. Tuberculosis has infected more than one and a half billion people in the world, about a quarter of the world's population, but has the "dishonor" of being associated with homelessness.[11] Lung illnesses, as a whole, have been unfairly categorized as dirty diseases, and the afflicted as unworthy of our attention. Ignored, underfunded, and forgotten: this is the medical history of lung disease.

The neglect has had serious consequences. Respiratory diseases, which include asthma and COPD, are among the top three leading causes of death both in the United States and worldwide. In America, these lung conditions traditionally trailed heart disease, cancer, and cerebrovascular disease as causes of death; however between 1980 and 2014, according to the Centers for Disease Control and Prevention (CDC), heart disease decreased by 59 percent, stroke by 58 percent, and cancer deaths by 24 percent, while chronic lower respiratory diseases increased by 40 percent.[12] The data is even more alarming for the period between 1965 and 1998, during which death rates from COPD increased by an enormous 163 percent, even as all-cause mortality declined by 7 percent.[13] In 2008, respiratory diseases in the United States for the first time replaced stroke as the third-deadliest disease, and it has kept that place ever since.

These statistics on the exploding burden of lung disease, dismal as they are, would be welcome in many other countries. Respiratory infections are the leading cause of death in low-income countries, where infants and children under five make up a disproportionate share of the four million deaths each year.[14,15] Globally, toxic indoor and outdoor air pollution are an issue for three billion people, and together these problems are responsible for eight million premature deaths each year. Ninety-one percent of people in the world live in places where air quality fails to meet World Health Organization standards.[16] All of these statistics point to a significant international health crisis.

Lung diseases show no signs of abating, nor do the persistent, alarmingly high smoking rates or the worsening air quality driven by climate change and pollution. Even more worrying, crises that pose a threat to the breath and lungs have grabbed headlines recently, from the lethal wildfires in California, the Amazon, and Australia, to the strange respiratory illness from vaping, to of course the devastating 2020-coronavirus outbreak that shut down the global economy and killed hundreds of thousands. These catastrophes show that we have not taken potential threats to our air seriously enough.

In the face of these challenges, some innovative doctors, scientists, and advocates are working extraordinarily hard to prevent and cure lung diseases. Given what we now know about genetics, biology, and medicine, there is no better time in history to be on the front lines of that fight—or, if you must, to be a patient with lung disease. The stories in this book illustrate the uniqueness of our current moment, showing how we've gotten to where we are with the lungs, and also pointing the way to a bright future.

PART I

THE PAST:
THE LUNGS SHAPED
OUR BEGINNINGS,
PHYSICALLY AND
SPIRITUALLY

Chapter 1

Oxygen, Then Existence

The story of our need for breath goes back many millions of years. As each person's biological life has a conception, gestation, and early, middle, and late stage, the same is true of Earth itself. Just as an infant coming into the world can flourish only when it has mastered breathing, Earth started flourishing only when some kind of breathing and oxygen use began.

The Earth has not always had oxygen in its atmosphere. Its early gases would have been toxic to most of the species alive today. But when oxygen first appeared, it changed the world radically. And, remarkably, we didn't know how oxygen first came to envelop our planet until the 1970s.

The universe, namely all matter that we see before us in the form of stars and planets and everything else contained within their apparent space, is thought to have emerged some fourteen billion years ago. Almost certainly, in the single instant of the Big Bang's explosion, the entire past and present matter of the universe burst into space and spread over the cosmos. Over time, various parts of the universe have expanded and cooled, with different solar systems springing up as stars exploded into violent supernovas and the leftover nebulae of gases condensed into solid matter.[1]

Our own solar system formed about 4.5 billion years ago. The other planets near us are basically rocky masses, but Earth is obviously different. Pictured from outer space, it appears as the aptly named Blue Planet, a cool, serene mixture of deep aqua oceans and swirling white atmosphere. It stands in stark contrast to the harshness of neighboring Mars, the Red Planet, or our own moon, white and barren.

But Earth came into being devoid of its beautiful oceans, lush green landscapes, and the give-and-take of evolution, life, and death. For the first four billion years of its existence, Earth fluctuated between extremes of heat and cold, its atmosphere a toxic mixture of nitrogen and carbon dioxide. And for the first two billion years of its existence, it had absolutely no oxygen in its atmosphere.

Oxygen is so important because of its ability to generate energy efficiently. Organisms derive energy from molecules called adenosine triphosphate (ATP), which are formed through cellular respiration. Without oxygen, cells, through a process called anaerobic fermentation, can still produce ATP—but only a measly two units from each molecule of sugar. This is highly inefficient compared to metabolism with oxygen, through which cells can produce thirty-six ATP units from each sugar molecule. Equipped with these extra units of energy, organisms are able to grow bigger, run faster, and jump higher. Without oxygen, the only living mobile organisms would be anaerobes, tiny creatures that are no match for this world's oxygen consumers.

Thus, for the first few billion years of its existence, Earth contained no plants and no animals. Oceans formed shortly after Earth came into existence, as the planet cooled and atmospheric water vapor condensed, but the only life that they could sustain were small, single-celled anaerobic microorganisms. Then, about 2.5 billion years ago, oxygen slowly began to be deposited in the atmosphere. It took a long time to reach a level of significance, but finally, about a billion years ago, the oxygen sinks of the Earth, mostly iron deposited in rock, became saturated. Oxygen then began to build up in the atmosphere and in the oceans. Termed the "Great Oxygenation Event," or GOE, this watershed precipitated an explosion of life, with ocean plants arriving about six hundred million years ago, and then later sponges, mollusks, fish, and finally terrestrial plants and advanced life.[2]

A single question remained for a long time, however: Where had all this oxygen come from? Something substantial must have happened

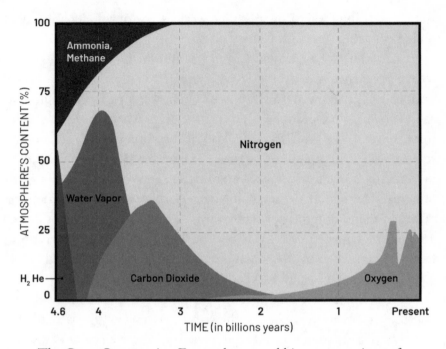

The Great Oxygenation Event: the natural history, over time, of atmospheric gases.

for a whole new gas to transform the planet in such a unique way. The story of how we began to understand where oxygen came from, and how it changed the world, is an extraordinary tale of hard work, keen observation, and luck (a combination that likely describes many, if not most, scientific discoveries). It's also a story that simply is not well known—but should be.

John Waterbury grew up in the Hudson Valley of New York but spent his summers in the coastal Cape Cod town of Wellfleet, Massachusetts. There, in the early 1960s, Waterbury wandered around the expanse of dunes that stretched into long beaches and looked out into the blue-green ocean water of the Atlantic. Not satisfied to remain on the shore, he took to the ocean in his Lightning dinghy racer. Surrounded by salt water and rolling waves, he was filled with a sense of wonder as his boat glided over the waters off Cape Cod.[3]

Waterbury's first academic stop was at the University of Vermont, where he earned a degree in zoology in 1965. After graduation, his options were narrowed down to two. There was a research position at the Woods Hole Oceanographic Institution in Massachusetts, a mere forty miles up the Cape from his Wellfleet summer home. If he didn't stay in academics, the draft awaited him, with a possible tour of duty in Vietnam. Not surprisingly, Waterbury chose Woods Hole. He spent four years there, studying nitrifying bacteria, little organisms that digest nitrogen-containing matter. Afterward, he enrolled in a doctoral program at the University of California, Berkeley, and spent a few years in Paris. He returned to the Oceanographic Institution in Woods Hole in 1975, this time to stay. At Woods Hole, Waterbury discovered how Earth had changed from a planet without oxygen, inhabited only by microscopic organisms, to one with oxygen, teeming with all sizes of life.[4]

During his doctoral studies at Berkeley, Waterbury found his passion in cyanobacteria, microorganisms that were known to colonize fresh water. More commonly known as blue-green algae, these organisms have properties more like those of plants than of bacteria. Foremost among these unusual properties is the ability to photosynthesize—to turn carbon dioxide and water into oxygen and carbohydrates. But in the 1970s, cyanobacteria were mostly known to colonize only small fresh-water areas and were thought to have had a limited role in the Earth's process of oxygen production. They were not talked about outside of a small academic circle, and no mention of them appeared in major oceanography textbooks.

After his doctoral studies, Waterbury settled into his job as a research scientist at the Oceanographic Institution. A primary mission in the field at the time was to study ocean bacteria, of which not much was known. Field trips were a regular part of the investigation, and in August 1977 Waterbury headed out on the research vessel *Atlantis II* to the Arabian Sea, the mass of ocean between India and Saudi Arabia known for having very high levels of inorganic nutrients and a rich marine life. His team's mission was to analyze samples from the ocean using a new technology: epifluorescence microscopy. The goal was to establish typical levels of known bacteria in the ocean with this new technique.

The basics of epifluorescence microscopy are straightforward. Tags, made up of the building blocks of DNA, are added to a sample of water, where they attach themselves to corresponding parts of the DNA of bacteria, like puzzle pieces fitting together. Under the blue light of the microscope, these bacteria then fluoresce green from their newly attached tags. If no matching bacteria are present, the tags won't be activated, and the view in the microscope will remain blank.

Before adding his DNA tags to the Arabian Sea water, Waterbury did one thing that all students are taught in science class, a mandatory step for every experiment at every level of science, from middle school classrooms to Nobel Prize–winning labs: he set up a rigorous control to ensure his results would be valid. Scientists know that controls are the backbone of all discovery. To find something abnormal, one needs to be able to see, and prove, the existence of what one thinks is normal. So, prior to adding the DNA tags, Waterbury analyzed an unaltered specimen of water from the Arabian Sea under the new epifluorescence microscope so that he would have a baseline for comparison.

Waterbury assumed he would see nothing unusual in the Arabian Sea water, but instead he was stunned. The blue light of the epifluorescence microscope went through the water, and a bright-orange fluorescence came shooting back out of the eyepiece. Because of his background in cyanobacteria, Waterbury recognized the orange light as the natural fluorescence of phycoerythrin, a photosynthetic pigment that works with chlorophyll to drive the all-important carbon-dioxide-to-oxygen-and-carbon reaction, making life on this planet possible. Cyanobacteria had never been reported to exist in deep-sea salt water, so this was a monumental new finding.

The initial discovery of cyanobacteria in the Arabian Sea was an introduction, but in order to be able to study saltwater cyanobacteria in depth, Waterbury knew he would have to grow them in culture. He tried for months, each time using a new medium and different nutrition to coax the cyanobacteria to replicate. But each time the same thing happened—within twenty-four hours the cells were all dead. Culturing these bacteria was a must if the study of saltwater cyanobacteria was going to advance. In order to succeed, Waterbury had to go back to basic environmental biology.

Ocean organisms and freshwater organisms behave very differently. Normally we think of ocean creatures as hardy and adaptable, and the ocean as a rough and wild place. Freshwater bodies, by comparison, seem quiet and tranquil, without sharks and stingrays and deadly jelly-fish. This is the human perspective. From the perspective of bacteria, the reverse is true.

The environments of freshwater bacteria and ocean bacteria are strikingly different. In inland bodies of fresh water, the temperature can vary widely, as can the amounts of nutrients and minerals. The summer and winter also produce very different living conditions in freshwater environments, which often host very different species depending on the season. By contrast, the ocean environment is exceptionally stable. The temperature variations are much smaller than they are in inland bodies of water, and the microenvironment of nutrients much steadier. Bacteria that thrive in a freshwater environment are what oceanographic scientists call "eutrophs," organisms that can handle an abundance of nutrients and highly variable temperatures. Saltwater bacteria, "oligo-trophs," require lower levels of basic nutrients. So, somewhat contrary to intuition, saltwater bacteria are more sensitive, more fragile than their freshwater cousins.

Over the course of a vexing year, Waterbury came to understand this. He fastidiously scrubbed all his culture flasks and test tubes, making sure not even a microscopic amount of calcium or other substance was left over. He then calibrated his culture medium to precisely reflect the nano-amounts of nutrients he had measured in the ocean water. Finally, after a year of meticulous work, and much to Waterbury's delight, cya-nobacterium from the ocean started growing for the first time outside of their natural habitat. The discovery of the species *Synechococcus* was official.

The questions then remained: How much of this stuff is out there, and what is its habitat? From the end of a wooden dock in Woods Hole, Waterbury filled a few jars with salt water, a little murky but otherwise unremarkable. He looked at the specimen under his epifluorescence microscope; it was teeming with cyanobacteria.

The study of cyanobacteria exploded over the next ten years. Hundreds of different species were identified in almost every ocean habitat on Earth. We now know that blue-green algae inhabit any body of water that is warmer than five degrees Celsius, usually in massive numbers, so massive that Waterbury refers to them as "those little beasts."

With their sheer numbers and diverse habitats, cyanobacteria are today recognized as the creatures primarily responsible for putting oxygen in our atmosphere. They do so through photosynthesis, the process by which plants, algae, and cyanobacteria capture sunlight and turn it into energy. The primary molecule that captures the sunlight is chlorophyll, which uses the energy from photons of light to drive the reaction of carbon dioxide and water to glucose and oxygen. The photosynthetic reaction also gives off energy that helps cyanobacteria convert atmospheric carbon dioxide to edible carbon, which is consumed initially by lower life forms but then carried up the food chain. This process makes cyanobacteria the source of a great deal of the food production on Earth. They are also responsible for a majority of Earth's oil, natural gas, and coal, all of which derived from settled matter (dead cyanobacteria) that condensed at the bottom of the ocean over millions of years. Indeed, cyanobacteria as a group are the most abundant species on Earth, and one of the most important for the purpose of life.

We tend to associate the process of photosynthesis with plants, but almost certainly cyanobacteria did this first. It is thought that, millions of years ago, ancestral cyanobacteria paired with larger cells, in a process called endosymbiosis, and evolved to become chlorophyll-containing chloroplasts, which allowed the larger cells to perform photosynthesis.

These chloroplast-containing cells eventually bound together to make the forerunners of present-day plants and algae.

The mastery of photosynthesis by cyanobacteria, and later by plants, is something, despite all our technological progress, we can still only marvel at. Humans figured out early how to burn carbon, but we still have not been able to produce it ourselves from carbon dioxide and light. If photosynthesis could ever be simulated artificially, it could be the golden key to solving our energy-production problems; it would also solve the problem of global warming, by making it possible to take carbon dioxide out of the atmosphere.

Looking back, we now know that the explosion of life during the Cambrian Period, some five hundred million years ago, was significantly fueled by rising oxygen levels in the atmosphere caused by oxygen production from cyanobacteria.[5] Higher animal life forms would simply not be here without these little creatures, nor would most plant life forms.

Our lungs developed to utilize oxygen and efficiently drive our metabolic reactions. We are aerobic creatures, and if the lungs are our most important organ, then oxygen is the most important gas in the atmosphere. Anaerobes exist, but they are constrained by an inefficient method of energy production. With oxygen, the possibilities of the world opened up. Almost every living creature on Earth is reliant on some method of oxygen extraction, and John Waterbury and others in the field of ocean bacteria helped show us where all this life came from.

With the existence of the new gas in Earth's atmosphere, the last five hundred million years of this planet have been radically different from the first four billion. The first period was marked by an absence of life, the second period by an abundance of it. The timing of those two appearances, of oxygen and life, is no accident. Oxygen is the life force, the source of life's infinite possibilities.

Along with the rise in oxygen from cyanobacteria, plant life began to flourish around this time. It first occurred in the ocean, and then inexorably these plant forms made their way onto the scorched orange land mass that, at the time, was completely devoid of anything except rock. First shallow moss colonized the rock, and then slowly more advanced plant life established itself. Trees came later, which further increased oxygen levels.

In the oxygenated ocean, animal life became increasingly sophisticated. With more plants came more oxygen, and with that, worms, mollusk-like clams, and jellyfish came into being, using primitive gills or simple diffusion to extract oxygen from the ocean. Eventually, over the course of tens of millions of years, creatures made their way onto the land that had been colonized by plants. Insects, spiders, and worms were the first to take advantage of a nascent verdant landscape. But they couldn't have made this remarkable transition without some kind of ability to utilize oxygen.

Worms have no functioning respiratory system. They derive oxygen from the moist soil around them, letting it dissolve through their skin and into their blood. Dry a worm out and you suffocate it. Spiders and insects have a respiratory system, but it is simply a long pipe, going through their bodies, that allows oxygen to diffuse into the surrounding tissues. In all of these species, there is no muscular system to augment the utilization of oxygen, and no way to increase the supply of oxygen significantly in a time of need. These primitive systems are limited by their lack of efficiency. This limitation prevents the bodies and brains of these creatures from growing bigger. They are constrained by their lack of lungs.

As worms and spiders were creeping out of the sea, life was advancing at a much faster rate in the ocean than on land. Creatures were both growing bigger and developing more complex organs. Vertebrate animals, with an endoskeleton and a skin covering, evolved, as did fish with familiar organs like a brain, liver, heart, and alimentary canal. These sophisticated vertebrates began to colonize many different aquatic niches, from the highest river streams to the deepest ocean trenches. The Devonian Period, spanning from 420 million to 359 million years ago, is known as the age of fish for the explosion in the number of species and in the number of habitats they colonized.[6]

Fish likely diversified because they developed an ability to utilize oxygen through an efficient circulatory system. A large part of that system is the gill. Most fish have a single slit on either side that allows water to pass through. As the water flows in, a vast network of capillaries embedded in the gills extracts oxygen from the water. The capillaries also expel carbon dioxide, in a gas exchange system parallel to our own. Most fish also have muscles around their gills that can cause them to flap and increase the stream of water and oxygen into the system as energy needs increase. It's a worthy system of oxygen utilization, and it explains why fish have developed into some of the biggest creatures on Earth.

In time, and only after they developed lungs as a way to extract oxygen from the atmosphere, fish made their way onto land. It is a single miraculous development, albeit spread out over tens of millions of years. It fascinates us because we can think of it as the moment of our birth, a symbol of when life as we know it was finally within reach. What made this transition possible was the creation of lungs, the organ that defines us as terrestrial creatures.

The metamorphosis of fish is thought to have begun in the shallow muddy waters where the ocean and land meet. There was a clear adaptive benefit to being able to stay out of the water for extended periods of time to take advantage of a landmass full of food in the form of plant life.

Exactly how lungs first developed in fish is a question that has long been debated. One thing that appears clear, though not intuitive, is that our modern lungs did not evolve from gills. Interestingly, the gills of some fish, most notably the walking catfish, have evolved into a partial lung. Native to Asia, but now taking over Florida, these fish have developed a very small area of gas exchange that opens only when they close their rear gills.

Our lungs, however, likely started as an outpouching of the esophagus as fish began to breathe by simply swallowing air that then diffused into the circulation by simple osmosis. Some fish have retained this early outpouching, known as a swim bladder, which is filled with air. Modern fish use the swim bladder as a ballast mechanism for buoyancy. But the bladder in some earlier fish developed into the lungs we know today.

One other important transformation necessary for fish to thrive on land was the development of legs to maximize maneuverability outside of water. Creatures with four appendages are referred to as tetrapods, a class that today is made up of all the mammals, reptiles, birds (wings count), and amphibians. Most likely, during the Devonian Period, about four hundred million years ago, the first type of tetrapod emerged from the ocean with newly, and simultaneously, evolved lungs and legs.

The fossil record from that time reveals clear signs that some fish were making attempts at coming onto land. These early colonizers had a more defined bony structure in their fins, and the beginnings of a lung along with their gills. One such fish was the coelacanth, which was thought to have gone extinct millions of years ago. This belief changed by chance on a sunny day in 1938, when a young woman in South Africa spotted something unusual on a fishing vessel, spawning an extraordinary fish story and an international sensation.

Marjorie Courtenay-Latimer was a museum curator from East London, South Africa, which lies between Cape Town and Durban on South Africa's eastern coastline. As part of her job, Marjorie received calls from fishermen coming in from the local waters with an interesting catch. The call that would change her life came on December 22, 1938, from Captain Hendrick Goosen, who had been fishing the waters of the Indian Ocean at the mouth of the Chalumna River. Marjorie came down to inspect the catch for any standout specimens and noticed a blue fin protruding beneath a pile of rays and sharks on the deck. Pushing the other fish aside, she came upon, as she would later write, "the most beautiful fish I had ever seen, five feet long, and a pale mauve blue with iridescent silver markings. It was covered in hard scales, and it had four limb-like fins and a strange puppy dog tail."[7]

Marjorie had never seen a fish like this before, and she sent a telegram with a crude drawing to Dr. James Smith, a local chemistry professor with a reputation as an amateur ichthyologist. Dr. Smith immediately saw the importance of this find and cabled back: "MOST IMPORTANT: PRESERVE SKELETON AND GILLS [OF] FISH

DESCRIBED." Because of his excitement, he cut two days off his vacation and went to East London, where he immediately identified the fish as a coelacanth, a ghost from the evolutionary past believed to have been extinct for sixty-six million years. It was named *Latimeria chalumnae* (from Marjorie's last name and the name of the river it was caught in), and from studying it, along with another one caught a few years later, scientists clearly saw from its anatomy that the fish represented an early transition from the ocean to land. First, it had a structure in the thorax that could be described as a lung, only in the coelacanth it was filled with fat. Second, its four fins had cartilage in them, unlike the simple fins of modern fish, making them clear forerunners of our modern limbs. Being a bottom dweller, the coelacanth used its fins in sequence for crude locomotion on the ocean floor.

The coelacanth caused an international sensation when it was "discovered" in 1938, but other species live among us that illuminate even more clearly the early development of lungs and legs. While the coelacanth has the beginnings of a lung, some fish have actual lungs. The most recognizable of these creatures are the mudskippers, three-and-a-half-inch fishlike creatures whose natural habitat is the muddy flats in the eastern part of Madagascar, as well as in parts of southern China and northern Australia. The beauty of the mudskipper lies not in its looks; in fact, its bulbous, puffy face and bulging eyes are naturally repulsive, its slimy body is off-putting, and its two fins, strangely placed on its back, appear pasted on in random fashion. But there is existential redemption for the mudskipper, because it has the remarkable ability to breathe in the water and on land. One minute it is happily swimming under water, and the next it's jumping onto the land, aggressively defending its territory with mouth gaping open and fins aggressively flared out. To be able to do this, the mudskipper has retained its gills, but has also adapted to absorb oxygen through its skin, its mouth, and the lining of its pharynx (the area below the mouth but above the esophagus and trachea). It can stay above water for days, sequestering its gills and keeping them moist under a flap of retractable skin. It has also developed rudimentary forelimbs—small arms that can push its slimy body around over its muddy habitat.

The mudskipper is not the only species to survive from this period of water-to-land transition four hundred million years ago. Amphibians, notably frogs, toads, and newts, can breathe using cutaneous respiration, in which blood passing past the skin picks up oxygen and releases carbon dioxide. Amphibians utilize this system both under water and on land. The Australian lungfish is another whisper from our evolutionary past. It is one of six remaining lungfish species, and the one that most effectively still straddles the worlds of the ocean and the air. Its demeanor is nonthreatening, and it has a long, olive-green, heavy snakelike body, small eyes, and four fins that help with propulsion both in water and on land. Size-wise it is not diminutive, averaging a healthy twenty pounds and measuring four feet in length. It inhabits the shallow, muddy fresh waters of Queensland, in northern Australia, an isolated, quiet place of sequestered species, seemingly frozen in time. At 370 million years old, the Australian lungfish also gives off the scent of the prehistoric, as if it would be at home dodging the bite of a pterodactyl or the sweeping jaws of a crocodile.

The lungfish's use of oxygen is impressive, as it can alternate between what a fish would do under water and what a land creature would do above. Unlike the mudskipper, the lungfish has a proper lung, with appropriate gas exchange units, not just the simple diffusion of air through a membrane. It can live several days above the water, where it feeds off plants that otherwise would be inaccessible. The lung also comes in handy when the water in the fish's natural swamp habitat runs low.

What the coelacanth, the mudskipper, and the Australian lungfish offer us is a fascinating window into our past, one that shows us how species experimented with different modes of oxygen extraction. Without oxygen and a mode of extraction, we would not exist, nor would most of the species around us.

The intersection of our existence, oxygen, and the breath is interesting not just as a story, but as a roadmap that points us in the direction of the future. Prominent scientists have warned us that life on this planet is fragile, that at any moment an asteroid or nuclear war could wipe us all out. They warn that the fate of mankind, and indeed of all species, may someday rest on our ability to get off the planet.

In order to do so, we must of course accommodate the lungs. We are now struggling again, some four hundred million years later, with the challenge, successfully confronted by the mudskipper, coelacanth, and lungfish, of learning to survive in an inhospitable environment. Unfortunately, we can't change our organ of energy extraction as they did, but we can try to change a toxic atmosphere to one that is more hospitable.

The first planet to be considered for colonizing is Mars, and the engineering process of making that planet's atmosphere hospitable is called terraforming. Numerous obstacles are present, including the extreme low temperature and lack of gravity compared to the Earth. An even bigger issue is the atmosphere of Mars, which consists of 95 percent carbon dioxide, 2.7 percent nitrogen, 1.6 percent argon, and only 0.13 percent oxygen. The air is also extremely thin, about one hundred times less dense than Earth's. So somehow we are going to have to make the atmosphere more dense, and fill it up with oxygen.

One plan NASA is developing is called the Mars Oxygen In-Situ Resource Utilization Experiment, or MOXIE for short. The idea is to produce oxygen from carbon dioxide, much as a tree does, by using electricity to drive a reaction of carbon dioxide into oxygen. The design is already in place to put a small version of a MOXIE machine onto a land rover and send it to Mars, where it will be tracked to make sure it functions properly. The hope would then be to build a much bigger version, which would help create oxygen for fuel as well as for the atmosphere.[8]

Another idea for putting oxygen into the environment of Mars is to set up biodomes throughout the planet, and then bring microbes from Earth to do what they have been doing here for millions of years. The best candidate would likely be a species of cyanobacteria, and one that already lives under extreme conditions here. There is plenty of nitrogen, their natural fuel, on Mars for them to utilize. The biodomes would then be monitored for oxygen production, and if the experiment is successful, many more of the structures could be built.[9]

In order for this manufactured oxygen to remain close to the planet, a denser atmosphere will have to be created. Scientists think creating a magnetic sphere around the planet, a protective coat of electromagnetic waves much like the one that surrounds the Earth, will keep destructive radiation from the sun away and minimize the effect of solar wind. A physical shield emitting protective magnetic waves will have to be strategically placed between the sun and Mars. If successful, it will allow existing carbon dioxide and newly created oxygen to build up, both warming the planet and allowing air pressure to increase. The hope is that this will help melt the ice currently trapped in the polar caps of Mars, unleashing water onto the planet once again.

This may all sound like science fiction, but there is a reasonable expectation that terraforming will be successful, and within a few hundred years we may be able to live permanently on Mars. The problems are the atmosphere, lungs, and breathing, problems that have existed since the beginning of terrestrial life. The first time around, these issues were resolved by evolution; this time, engineering is needed.

Chapter 2

We Must Inhale and Exhale.
But Why?

Sometime after midnight, one deep January night, I remember walking into my bedroom and staring down at my first child in her crib, born on New Year's Eve and now barely two weeks old. The silver moonlight poured in through the window, illuminating her outline. Her eyes were closed tight, her arms thrown over her head as if in an eternal stretch, her head tilted slightly to the right. Her intoxicating newborn smell brought rapture and calm.

Like millions of parents before, I considered the serenity of a newborn's sleep, its restorative depth, but I also instinctively checked on function, to make sure life still blazed within despite the outward calm. This meant checking her belly, to make sure she was breathing. Of course she was: her chest and abdomen moved up and down under her blanket in the rhythmic flow we all recognize as life.

When we observe our loved ones sleeping, old or young, human or pet, we are instinctively drawn to their breath. There is something essential in it we are all attuned to, something we both automatically and unconsciously equate with life. Each time we check on each other, we are validating the words of the Roman philosopher Cicero, *dum spiro, spero*, "As I breathe, I hope."[1]

Physiologically, what we are observing is the miracle of gas exchange, taking from the atmosphere an invisible element and bringing it into our body to be consumed. The process begins with a signal from the brainstem, the primitive part of the brain at the base of the skull, which travels through nerves down to the muscles of inspiration, instructing them to contract. The biggest and most important of these muscles is the diaphragm, a thin, dome-shaped sheet of skeletal muscle that separates the thorax (chest cavity) from the abdomen.

With each signal, the diaphragm contracts downward, pulling the thoracic cavity and the lungs with it. This creates negative pressure in the trachea and lung tissue, and with this negative pressure air rushes in, just as water flows down a river. Entering through the mouth or nose, the air travels down the back of our throat, past the vocal cords, and into the trachea. About halfway down the sternum, the trachea divides into the left and right bronchi, which divide again and again into lesser bronchi, termed bronchioles. The air moves through the bronchioles, which extend deep into the lungs, like tendrils from a star erupting in space, until finally it penetrates into cave-like recesses, deep in the lungs. Known as alveoli, and resembling cells of a honeycomb, these grapelike clusters at the end of the increasingly narrow breathing tubes are where gas exchange occurs.

Continuing its natural flow from an area of high concentration to one of low concentration, oxygen moves effortlessly through the thin surface of alveoli, just a single cell thick, to the adjoining capillaries. Here, thousands of hungry red blood cells grab oxygen, and together they are pumped by the heart to the arteries and then to the tissues of the organs, which are infiltrated with a vast network of capillaries. At the tissue level, oxygen hops off the red blood cell and diffuses through the capillaries into the cells of whatever organ or muscle is nearby.

Within each cell are mitochondria, the specialized organelles in which cellular respiration takes place: oxygen joins with glucose to produce carbon dioxide, water, and adenosine triphosphate (ATP). ATP is our primary source of energy, the molecule that drives many of our bodily reactions, including muscle contraction, enzyme production, and the movement of molecules within our cells. ATP causes these reactions by breaking off one of its phosphate groups, whose electrons are in a high-intensity state, and transferring this energy to drive the necessary processes of the cell. Now adenosine *di*phosphate, it gets recycled back

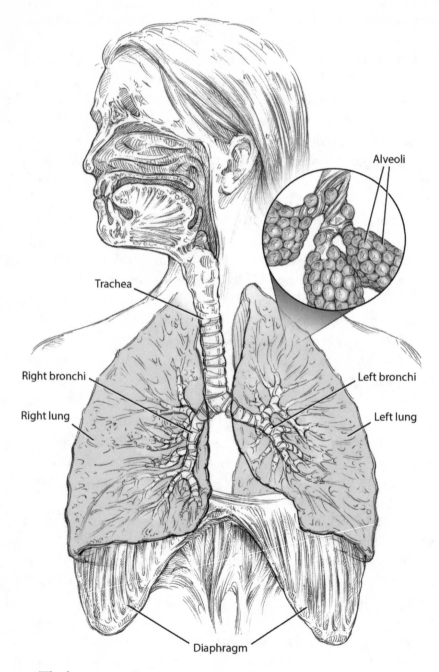

The human respiratory system.

to the mitochondria to become high-energy triphosphate again through the ongoing process of cellular respiration.

A byproduct of oxygen use and cellular respiration is carbon dioxide (CO_2), which diffuses out of the cell, into the blood, and back into the capillaries, which now flow to veins. Not used by our body, CO_2 is shunted back to the lungs by our venous system, where it diffuses into the alveoli. From there, the air, with its new mixture of gasses, is pushed out through the network of bronchioles and bronchi as the diaphragm relaxes with exhalation, and ultimately expelled out of the mouth or nose and back into the atmosphere. The CO_2 disperses easily into the air, where its levels are very low at 0.04 percent of atmospheric gas. Levels of oxygen in the atmosphere stand at a comparatively robust 21 percent, so on our next breath we are able to fill up again with this molecule of life. (The rest of the atmosphere is almost all nitrogen, harmless to us but also useless.)

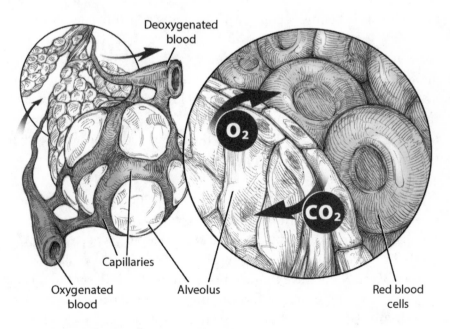

Gas exchange at the alveolar level.

We are attuned so closely to this process of inhalation and exhalation in our slumbering loved ones because we have an instinctive understanding of its urgency: while we can skip meals, breathing must be continuous. The system must be finely coordinated as gas levels need to be kept within a very narrow range in our blood. Receptors in our aorta and carotid artery continuously monitor levels of oxygen and carbon dioxide and send feedback to the respiratory center in our brainstem. Even the slightest change in the levels of gas will trigger more signals or fewer to go out to our muscles of inspiration. The activity in the respiratory center of our brainstem also feeds back to our cerebral cortex, or higher brain, making us aware of any impending danger. This creates the alarming sensation we are all familiar with if our brain senses something wrong with our oxygen or carbon dioxide levels, as when we hold our breath.

Carbon dioxide is what causes most of the initial problems when we hold our breath, because as it builds up in our blood during a breath hold, it begins to break down into acid. This acid is toxic to our cells, especially as it begins binding with proteins and other molecules it shouldn't, impeding the normal function of cells. As the breath hold continues, lack of oxygen also becomes a problem, and as cellular respiration in our mitochondria ceases with the dearth of oxygen, cellular death ensues. The cells of the heart muscles are especially sensitive to this, and cardiac arrhythmias can ensue in extreme cases of too much carbon dioxide or too little oxygen. Breathing is the most important thing we are aware of doing, and the body regulates it tightly.

The foundation of our understanding of these pulmonary processes, and indeed of all Western medicine, was laid in ancient Greece. After the mythical Apollo and his mystical son Asclepius, whose rod is the symbol of medicine today, the first legendary but real figure in the history of medicine was Hippocrates, born in 460 BCE on the Greek island of Kos. He is forever known for creating the oath that all physicians still recite when they receive their diploma, and is deservedly known as the Father of Medicine for his insight that disease was the consequence of natural processes and not the work of magic or the gods.[2]

In addition to contemplating many other anatomical systems, Hippocrates studied breathing. He recognized that the inhalation of air was fundamental to life. For this reason, Hippocrates and the Greeks viewed air as something vital and transcendent. They called it *pneuma*, which literally means air or breath but for the ancient Greeks also meant life force. This *pneuma* was inhaled and then passed through the lungs, into the blood, and on to the heart, where it became the *pneuma zoticon*, or vital spirit. This vital spirit was then carried to the organs, including the liver and brain, where it was transformed into *pneuma psychicon*, or animal spirit, which was considered the driving force, created by the body from air. From spirit (air or *pneuma*) to vital spirit to animal spirit, the Greeks and Hippocrates insightfully saw the essence of our existence as a continuum from the atmosphere.[3]

Some five hundred years after Hippocrates, Claudius Galenus was the next great figure to change how we think about breathing and circulation. Better known simply as Galen, he was born in September 129 CE, in the Aegean Sea town of Pergamon, part of modern-day Turkey. His father, a wealthy patrician, originally had plans for his son to become a philosopher and statesman. These plans changed when the father dreamed that the mythical physician Asclepius visited him with a decree that his son study medicine. Galen's father spared no expense, and Galen was educated at the best institutions throughout the Roman Empire.

When he finished his studies, Galen settled into practice in Pergamon. He became the personal physician to the gladiators of the high priest of Asia by performing an act of daring. According to his own report, he eviscerated a monkey and then challenged the other physicians to repair the damage. When none stepped forward, he did the surgery himself, successfully restoring the monkey and winning over the high priest. He later moved to Rome and became the personal physician to several emperors, most notably Commodus, who reigned from 161 to 192 CE.

Galen contributed to many areas of medicine, and added to our understanding of the lungs and circulatory system. He noted that "blood passing through the lungs absorbed from the inhaled air the quality of heat, which it then carried into the left heart."[4] Human dissection was prohibited by Roman law, but Galen dissected both primates and pigs. He was the first to describe the two separate systems of circulation, the arteries and the veins. He believed that the liver, with its dark purplish interior, was where blood originated. From the liver, he postulated, half of the blood went out to the veins and was delivered to the tissues and consumed. The other half went to the lungs via a vein, where it picked up *pneuma*, then on to the heart, the arteries, and then the tissues.

Although Galen's theories on blood flow would be proven to be partially wrong, he was important, like Hippocrates, because of his methodology. Galen cemented the notion that medicine and disease were not the products of divine intervention from the gods but could be discerned from empirical evidence and deduction based on observation and cause and effect. Nonetheless, more than a thousand years passed before his ideas on the movement of oxygen within the blood's circulation were corrected.

In somewhat ironic contrast to his philosophy, Galen's ideas were accepted as gospel over the centuries, particularly his ideas on the flow of blood into both arteries and veins, and on the liver as the epicenter of blood production. Fortunately, the idea that the breath was important also did not change, as can be seen when the Renaissance scientist Alessandro Benedetti poetically wrote in 1497: "The lung changes the breath, as the liver changes the chyle, into food for the vital spirit."[5]

The man who changed our understanding of blood flow was William Harvey, an English physician trained in Padua, Italy. He, like Galen, had a big personality, and he frequently walked around with a dagger on his belt, as was the fashion in rough-and-tumble Renaissance Italy. His opinion of his fellow man was not high, and a biographer who lived at the same time claimed, "He was wont to say that man was but a great mischievous baboon."[6]

Upon establishing himself in England after his Italian internship, Harvey published *De Motu Cordis et Sanguinis* (*On the Motion of the Heart and Blood*) in 1628, cementing his reputation as a giant in the history of medicine. This work was seminal in our understanding of the basic physiological principles of how blood travels in the body. Harvey had two breakthrough insights. He noted that he had learned from his Italian mentor that the veins all had one-way valves, pointed away from the tissues and organs and toward the heart. Why the venous system, which was postulated by Galen to bring blood to the organs as the arteries did, had back valves to keep blood away from those parts of the body was not easily explained.

Harvey's second important observation was made through painstaking dissections of humans and animals. He calculated that the output of the heart was a lot greater than previously thought, about five liters per minute. He correctly reasoned that the tissues could not consume that volume of blood each minute, as proposed by Galen. He needed a more reasonable explanation, a system that was simple yet elegant—and that went against fifteen hundred years of dogma. So he proposed something often found in nature: a system of reuse and recycle, a continuously flowing system, a circuit, or, as we know it now, circulation. Blood is not consumed by the tissues—it is reused over and over.

As Harvey correctly deduced, blood moves in a circle. From the arteries, it goes past the tissues, where oxygen hops off and carbon dioxide hops onto our molecules of hemoglobin, then to the veins and the right side of the heart, through the pulmonary artery to the lungs, where the carbon dioxide produced from tissue respiration is released and oxygen is picked up, then to the left side of the heart, then out again through the vast arterial system and back to the tissues. Blood is continuously cycled in a beautiful loop, with the bone marrow (not the liver) replenishing the red and white cells as needed.

The idea was received as scientific heresy at first, thanks to the primacy of Galen's views. In response to the widespread doubt, Harvey gave a lecture in May 1636, which stands out for being as instructive as it was gruesome. Dressed in a billowy white dissecting gown, he spoke in

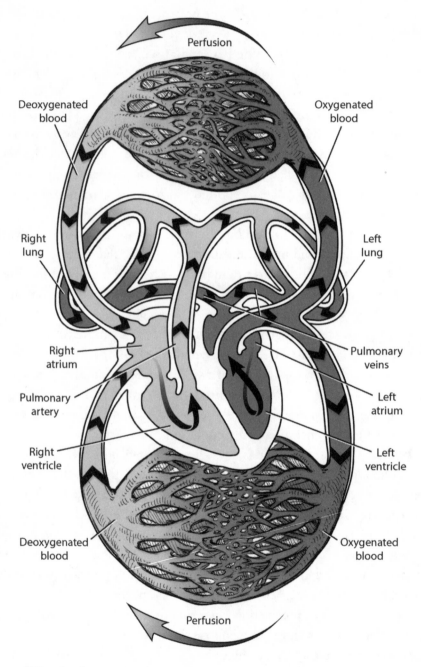

The circulatory system.

Latin to the professors, students, and general public at the University of Altdorf, in Bavaria, Germany. First, a live dog was placed on a dissecting table and strapped down. Next, Harvey declared, "It is obviously easier to observe the movement and function of the heart in living animals than in dead men." With this, he cut open the thorax of the writhing dog with a knife, to expose the beating heart, and then sliced the blood vessel next to the heart to show the audience the vast amount of blood now spurting out. His goal was to drive home his point that the heart was a pump, and that this amount of blood could not be consumed by tissues but must recirculate.[7]

Despite the theatrics, many of Harvey's contemporaries remained skeptical. Caspar Hoffman, a scientist present at his University of Altdorf lecture, declared, "*video sed non credo*," "I see it, but I don't believe it." Other critics pointed out two big leaps of faith Harvey had to make in order for his theory to work. The first was that there had to be some network of vessels between the arteries on one end and the veins on the other. We now know about capillaries, but Harvey had no tools to see or discover those tiny blood vessels. He made an educated assumption, as is so often needed in science. Confirmation would come just a short time later, in 1661, when Marcello Malpighi published his work *De Polmonibus observationes anatomicae* (*Anatomical Observations on the Lung*), confirming with the use of the microscope that capillaries did indeed exist.

The second leap of faith related to why and how the blood changes from dark to bright. Again, Harvey had a sense that something essential in the atmosphere, drawn in by the lungs, turned blue blood a bright scarlet. Nobody then had any idea about the role of oxygen, but Harvey intuited it in *Lectures on the Whole Anatomy* (1653), when he made the profound observation: "Life and respiration are complementary. There is nothing living which does not breathe nor anything breathing which does not live."

Establishing what was in the atmosphere that the lungs and body needed to capture would take a lot longer. The ancient Greeks had identified the air as one of the four classical elements, along with fire, water, and earth. Throughout the following centuries, air was believed to be a single substance. Not until the eighteenth century did scientists begin experiments to tease out the different chemical elements, including those contained in air. Joseph Priestley is credited as one of oxygen's first discoverers, his experiments taking place in 1774 and then being published over the next few years in *Experiments and Observations on Different Kinds of Air*. One of these experiments documented that both a mouse and the flame of a candle would die in a sealed jar of air. Next, Priestley created a new gas by focusing light with a magnifying glass–like apparatus onto a piece of mercuric oxide—and he noticed that this new gas kept both the candle lit and the mouse alive a lot longer than unadulterated air. Priestley shared his observations with French scientist Antoine Lavoisier, who would conduct further experiments on the purification of air—and give us the name *oxygen*.

Harvey's basic model of circulation remains intact today. We have improved our understanding of inflammation and genetics and cellular movement, but our ideas about what our circulation is doing and why have not changed. To break with centuries of established dogma was not easy, but an analysis of Harvey's methods reveals how most scientific breakthroughs are made.

First, Harvey ignored the dominant theory (in this case, one that had reigned for the previous fifteen hundred years). He then made a few astute observations and, based on limited data, posited a unique theory. He tested his hypothesis with more data collection and saw that it seemed to hold. He stuck with it despite two holes in his model (no knowledge of capillaries or oxygen). With this knowledge, we today understand the importance of keeping a patient's circulation moving with fresh oxygen. We are also painfully aware of just how severe the consequences are when this circuit is disrupted.

The brain recognizes the importance of the breath and guards respiration very closely by monitoring oxygen and carbon dioxide levels. To keep up with demands, the system was set up to tolerate a lot of failure, with five hundred million alveoli present in our lungs. Spread out, these alveoli would cover approximately one hundred square meters, the size of a tennis court. Because of this, it's possible to lose one lung entirely and still function adequately. Another fail-safe mechanism is the efficiency of gas transfer, with the surface that oxygen must pass through to get out of the alveoli and into the capillaries microscopically thin at one-third of a micron. A micron itself is a billionth of a meter, or a thousandth of a millimeter. The distance oxygen needs to traverse to get from the alveoli to the capillaries could double without any noticeable shortness of breath at rest.

Unfortunately, there are times when the system gets overwhelmed and needs help from technology. One night, when I was an intern at a hospital in Boston, I was tasked, despite my inexperience, with getting a man with greatly reduced lung function through the night. Fortunately, I had a lot of help from people who actually knew what they were doing. The practice of medicine can be as much an art as a science. This was one of those occasions.

Past midnight, I sat with my resident physician at the nurses' station, anxiously awaiting the arrival of Leonard Joseph, a patient from the deep woods of Maine who had gotten knocked down hard with an infectious pneumonia, which had caused a massive outpouring of inflammatory cells into his lungs, clogging up his alveoli so that gas exchange was greatly impaired. He had been placed on a ventilator, and even with this increased level of support, his lungs were stiff and struggled to expand. Oxygen was not getting into his body as he needed it to, and on the other end of cellular respiration carbon dioxide was not getting out.

When the doors finally swung open and the emergency medical technicians wheeled Mr. Joseph out of that cold January night, I nervously accepted a thick package of notes from the Maine hospital that had first seen him. I followed the resident into the intensive care unit room as the patient was being transferred into his new bed, his home for the next month. I saw he was getting 100 percent oxygen through the tube in his throat, a large amount compared to the 21 percent we normally get from the atmosphere.

We gauge how effective the transfer of oxygen to the blood is by measuring the pressure that oxygen produces within the blood of an artery, which is a reflection of how much oxygen is present in the blood. The radial artery, the easily accessible one that feeds the hand, is usually used to obtain a blood sample, which is then sent to the lab for analysis. Historically, the pressure of oxygen in the blood has been measured using a column of mercury, and gauging how much mercury the gas is able to move. A normal, healthy subject, breathing in a 21 percent oxygen mixture from the atmosphere, will generate a pressure of oxygen in arterial blood of about 95 millimeters of mercury (mmHg). The patient from Maine, however, had an oxygen level of only 60 mmHg, and this was when he was breathing in a 100 percent oxygen.

Once the level of oxygen in the blood drops below 60 mmHg, our tissues do not receive enough oxygen. This is when brain cells begin to die, and the heart becomes irritated. Clearly this was a critical situation, and one without an easy solution. Normally we would just turn up the oxygen level on the ventilator, but it was already at 100 percent. To stabilize this patient, something more creative was needed than simply turning up a dial.

The attending physician appeared a few minutes later, and he and the resident discussed some advanced maneuvers to help keep oxygen going in and carbon dioxide coming out. The last thing the attending suggested was "proning." Then he disappeared. The first question I asked was, "What's proning?" The resident, Kevin, looked at me wearily and said, "Proning is when you flip the patient over and have them lie on their stomach on the bed."

I still didn't quite get it. A groan followed. (There are many potential sources of humiliation in the hospital for an intern.) "'Why?' To help with oxygenation. And ventilation. You need to go learn about lung physiology. And you need to go read your *West*." Kevin shook his head and sat down to go through the mound of material in front of him.

The *West* my resident was referring to was a book by John B. West. John B. West is not known outside the small circle of lung medicine practitioners, but within it he stands tall, and through his research and textbooks he has cemented a reputation as the leading educator within all pulmonary medicine in the last hundred years.

Dr. West was born in Adelaide, Australia, in 1928 and developed an interest in science at an early age. He moved to England for his PhD, and then to the University of California, San Diego, to continue his research in pulmonary physiology. There he discovered some unique things about how the lungs work, specifically about how different areas of the lung can have very different blood flow and air flow. Later in his career, he wrote a textbook on pulmonary physiology that changed how we educate medical students. And even later, at the age of seventy, he changed his focus to the physiology of the bird lung.

Following Kevin's advice, I picked up an updated edition of West's 1974 textbook, *Pulmonary Physiology: The Essentials*. The volume is slim in the hand, and its two hundred pages of easy type and large diagrams belie the power of its contents—it is still the starting point for learning about modern pulmonary physiology for physicians today.

Dr. West begins his book with a review of the structure of the lung, and right away he makes an extremely important and powerful statement: The architecture of the lung follows its function. Everything about the lungs should be considered within the context of that fact: *form follows function.*

This statement is one that broadly defines an entire field of biology. The disciplines of comparative and evolutionary biology consider the origins and development of life around us in the context of form following function. The African lion, for example, lives in the grasslands and is almost strictly a meat eater. For that, it needs to be a predator. Its body is powerful and fast, but it can travel only in short bursts. It has big retractable claws and huge powerful teeth to take down its prey. Its niche, diet, body, teeth, and feet with claws all coalesce into one well-defined purpose. The wild dog, by contrast, has a more diverse diet, so its teeth include some canines but also some molars. The dog's legs and body are built for speed but also for distance, and it has no big claws, since they're not needed. Looking at the world from the form-follows-function perspective can be a powerful tool for analyzing nature's systems.

Dr. West states that the main function of the lungs is to facilitate gas exchange. First, oxygen needs to come into the blood and be carried to the cells of the body to keep metabolic processes going. Ventilation—the process of carbon dioxide being released from the bloodstream—needs to follow. Since form follows function, the body must have the ability to augment the work of the lungs to get appropriate amounts of oxygen into the blood and carbon dioxide out even when there are changes in our metabolic demands. Exercise, for example, is a state in which our tissues demand more oxygen and produce more carbon dioxide than usual. Illness from bacteria or viruses is another state that can increase our metabolic demands tremendously as a result of inflammation.

The lungs, for the most part, handle these demands with ease and flexibility. From a starting volume of five liters of air per minute at rest, we can increase our breathing to exchange ten, twenty, and even thirty liters of air per minute, an astonishing amount of gas. Our respiratory rate naturally increases in this situation, but the volume of air with each inhalation also increases, as muscles in our neck and our abdomen not usually used for breathing spring into action, helping to stretch our lungs to accept and release more air. These adjustments are important, since we must exist within a tight physiological space and maintain levels of oxygen and carbon dioxide within a narrow range. Our lungs, with help from the muscles in our chest wall, have the ability to keep us within that range under a wide variety of conditions. The trouble comes when something interferes with this powerful, but at times delicate, system.

Keeping his blood supplied with fresh oxygen, and his ventilation appropriate to expel enough carbon dioxide, was the problem Mr. Joseph was facing that dark January night in Boston. Well past midnight, I could feel the temperature in my body dropping as we prepped our patient for a procedure to put a large-bore intravenous line into his neck. He was on a lot of antibiotics, and unstable on a ventilator, so we deemed that he needed larger intravenous (IV) access to get medicines in more rapidly. For this, we needed to access the internal jugular vein in his neck, and as we carefully scrubbed his neck with cleaning solution, Kevin kept a running commentary.

Kevin spoke about how the patient was in ARDS [acute respiratory distress syndrome], so we would need to keep the pressures in his lung low and try to minimize the work of breathing. He discussed how we would have to keep a close eye on the patient's oxygenation, and to consider advanced oxygenation and ventilation methods—such as inhaled prostacyclin or maybe even ECMO [extracorporeal membrane oxygenation]—if we couldn't make some progress in the next few hours. Or perhaps proning.

I did know some of what he was referring to—that oxygen levels needed to be kept at least 60 mmHg in the blood. I also knew that ventilation is measured as the product of the number of breaths in a minute multiplied by how much air is moved with each breath (or tidal volume) and that we care about ventilation because it is the primary determinant of how much carbon dioxide is in the blood. When carbon dioxide builds up in the blood, it dissociates into free hydrogen molecules, which are essentially acid, the amount of which is measured on the pH scale.

The *pH* stands for *potential of Hydrogen*, and it is a direct measure of how many hydrogen molecules are present in a solution. The scale typically spans from 0 to 14, with 7 marking the exact middle of the scale. Water at 25 degrees Celsius has a pH of 7 and is considered completely neutral. If there are a lot of hydrogen molecules (H+), we call the solution acidic. On this side of the scale are drinks like black coffee, with a pH of around 5, and tomato juice, with a pH of around 4 (the pH scale is inverse, and a lower pH indicates more acid). On the other side of 7, we call solutions basic, and examples include baking soda–containing liquids, with a pH of 9, or ammonia, with a pH of 11. These solutions have far fewer hydrogen ions than acidic solutions do.

The pH of our blood is 7.40, and it must be kept in the very narrow range between 7.35 and 7.45, with 7.40 being optimal. Living within this pH range is extremely important, because the proteins of our cells—and subsequently metabolism itself—begin to break down when our pH goes too low or too high. The kidneys help to expel or hold onto acid as needed to adjust our pH, but the lungs are a far more powerful system for regulating pH through carbon dioxide, which, as mentioned, breaks down to acid in our blood. We regulate carbon dioxide and acid by simply increasing our breathing and ventilatory rate when too many hydrogen ions are present and the pH is too low, or by slowing it down and letting acid build up when the pH goes too high.

A classic example of when CO_2 rises is during exercise, and breathing increases concomitantly to expel the CO_2 so our pH doesn't drop too low. An opposite example is when we hyperventilate at rest without an increase in CO_2 production, as can happen during a panic attack. Here one is blowing off too much CO_2 and acid, and our pH climbs to dangerous levels, which is why one may be told to breath into a paper bag to inhale the expelled CO_2, restoring needed acid to the blood.

As for our patient from Maine, what I didn't know, but would soon learn, was what we were going to do if ventilation wasn't adequate and his pH was off, or if we couldn't maintain the appropriate level of oxygen in his blood. Meanwhile, I stood beside him with a large-bore needle, about to drive it deep into his neck to look for the internal jugular vein, when Kevin somewhat offhandedly mentioned, "Oh, and by the way, if you go in too deep and puncture his lung with that needle, he's probably going to have a cardiac arrest and die. So be careful, please." The top of the lung comes up very high in the chest cavity and lies just below the neck. I inserted the needle carefully.

The issue with our patient from Maine was definitely one of impaired gas exchange—that I understood—but my resident's term *ARDS* did not mean much to me at the time. The first known mentions of this mysterious lung disease in the medical literature were made in 1821 by the French physician René Laennec, who also invented the stethoscope. In his book *A Treatise on the Diseases of the Chest*, he described the death of a patient with lungs filled with water but without evidence of heart failure. Fluid in the lungs was certainly familiar to doctors at this time, but it was almost always preceded by the left side of the heart failing. The circulation in the chest begins as our veins empty blood into the right side of our heart, which pumps it to our lungs, where it travels to the left side of the heart to be pumped through the rest of the body. If the left chamber of the heart fails, as often happens with cardiac disease, the blood backs up into the lungs and fluid spills out into the alveoli. But Laennec noticed that in a certain subset of people, fluid was spilling into the lungs without heart failure, without high pressure. The capillaries of the lung were simply leaky, and patients were essentially drowning.

Later in the twentieth century, more of these cases started being reported, often when a soldier was being resuscitated after sustaining an injury in battle. Soldiers would get revived with blood and recover from their wounds, but strangely, afterward, their lungs failed, filling with fluid and later hardening up like a stone. Names like "DaNang lung" and "post-traumatic lung" sprang up in the literature. The disease itself was largely a mystery. It was also a devastating illness, with the mortality rate approaching 80 percent.

Part of the mystery was solved in 1967 in a paper by Dr. David Ashbaugh, of Ohio State University, and colleagues at universities in Denver and Michigan. Together they gathered information on a series of similar cases and coined the new term: *acute respiratory distress syndrome*, or *ARDS*. Their paper, published in the journal *Lancet*, describes the injuries and lung pathology of twelve patients who had a pattern of respiratory failure caused by too much fluid in the lungs.[8] The injuries the patients sustained before respiratory failure were disparate—some had trauma, others pneumonia, still others an inflamed pancreas. But their respiratory failure was similar—the extra fluid in their lungs, from leaking capillaries, would lead to dysfunctional inflammation and scarring, eventually turning their lungs rock-hard. As a result, all of these ARDS-afflicted lungs had trouble getting oxygen in and carbon dioxide out.

Dr. Ashbaugh's paper was a landmark study in that it put a name and a face to a mysterious condition, the first and often the most important step in successfully treating a disease. (One can't study a disease that has not been properly and accurately described.) And, quite extraordinarily, much of what is described in the paper is still relevant today. Unfortunately, the continued significance of this paper is also a sign of failure. According to the paper, ARDS is a disease in which some dramatic insult to the body occurs, which then causes lung inflammation and capillary leakage with no evidence of heart disease. The findings have not changed because the syndrome is stuck at the stage of description. There was no cure in 1967, and there is still no cure today.

So far, the efforts of physicians to slow or reverse the inflammation and leaky capillaries characteristic of ARDS have been futile. The journals are littered with descriptions of attempts involving different medicines—from steroids to inhaled nitric oxide to inhaled prostacyclin —that have come up short. Frequently, these treatments offered hope in mouse models, but every human drug trial has failed.

Even if no single medicine has been shown to improve outcomes, the mortality rate has dropped significantly, from the 80 percent in the 1960s to 40 percent today.[9] This improvement reflects a persistent focus on appropriate ventilator settings, nutrition, and physical therapy. There is much one can do in medicine without using drugs, and the dramatic decline in ARDS mortality speaks to this. People like John B. West, investigating how the lung moves air and blood, have been a big part of this progress. Still, with 10 percent of all medical intensive care unit admissions resulting from ARDS, it remains a formidable problem.

After we successfully put the central IV line into Mr. Joseph's neck, Kevin and I took a closer look at his ventilator. It had initially been quiet, but now its high-pressure alarms were starting to trigger loud pinging noises, indicating a failure to push air into the patient's very stiff lungs.

We called the respiratory therapist, and he and my resident came up with a plan to lower the driving pressure of air from the ventilator while giving the air a longer time to get in, easing the flow into Mr. Joseph's rigid lungs. We also had the nurse give him a paralyzing level of sedation, calming any involuntary fighting he was doing with his own respiratory muscles.

The modifications seemed to work, and the ventilator alarms went quiet. But this only told us that the machine was happy. The resident instructed me to check the level of gases (oxygen and carbon dioxide) in the patient's blood to see if his body was happy. A few minutes later the results were relayed from the lab. Our patient's oxygen level was just above 60 mmHg, and his carbon dioxide was around 48, corresponding to a pH of 7.30. These numbers were not great, but good enough to get him through what remained of the night.

For the rest of that month, I saw Mr. Joseph every morning at six o'clock, my first patient of the day. I would analyze how he was doing with oxygenation and ventilation, looking for any improvements. In the evening, at home, I would thumb through John B. West's book. By the end of the month, I began to understand some of the nuances of how gas exchange works, how different parts of the lung receive very different amounts of oxygen and blood flow. Specifically, the lower lobes generally get both a lot more inhaled air and more blood flow. Some of this is likely due to the effect of gravity.

With this knowledge of variable blood flow and air flow within the lungs, researchers eventually came up with the idea to minimize airflow in stiff lungs like Mr. Joseph's. The reasoning behind this idea is that, since blood flow and circulation are almost certainly compromised given the level of inflammation in the lungs, there is no need for a normal amount of air in each breath. Before this protocol was established, physicians had been blowing too much air into diseased lungs, creating too much stretch, and that extra stress was creating more inflammation. John B. West helped us appreciate that air flow and blood flow can be variable, and attempts should be made to match them. The breakthrough 2000 *New England Journal of Medicine* study showed that deaths were significantly fewer in ARDS patients who received less air when on the ventilator.[10] Practice in medical intensive care units all over the world changed overnight. No study before this, and no study since, has had such a dramatic impact on what we do in the medical intensive care unit.

Mr. Joseph was admitted to the hospital in January 2002, so this article was fresh in everybody's mind, and throughout the month we kept the air flow in Mr. Joseph's lungs to the absolute minimum possible while still maintaining ventilation. To compensate for a very low amount of air with each intake, we turned up his respiratory rate from the normal twelve breaths per minute to thirty, even thirty-four at times. Normally, a respiratory rate of thirty-four breaths per minute is not sustainable, but with a machine it is. We considered proning, or flipping Mr. Joseph onto his stomach, while he was on the ventilator, to further decrease stress on his lungs by reducing the effect of gravity, thus giving them a rest and a chance to heal. The front of the lungs is where alveoli are often less affected in diseases like ARDS. (Most recently, proning is commonly being used for patients with COVID-19 as the inflammation from pneumonia almost always begins in the lower part of the lungs.)

Throughout the month, Dr. West taught me about theory, while Mr. Joseph taught me about the real world. Slowly but surely he made progress, his stiff ARDS lungs loosening up enough that he could breathe on his own during the day, while using the ventilator at night. He eventually went to a rehabilitation facility, and from there, presumably, home to the wilds of Maine. As happens so often in the practice of medicine, all we did was keep him alive until he was able to heal himself.

Today, even though specific drug treatments for ARDS are nonexistent, other therapies have shown progress. Foremost among these is extracorporeal membrane oxygenation (ECMO), where blood is taken out of the body, run through a machine that removes carbon dioxide and adds oxygen, and then returned. It functions, in essence, like an artificial lung. It is not a long-term solution and only serves to buy time for the lungs to heal themselves. Studies of this treatment in adults with ARDS have had conflicting results, but it is an option for those who fail on the traditional ventilator.

Looking further into the future, the promise of stem cells is not just a dot on the horizon, but a viable therapy that is now making its way through clinical trials. Stem cells have the ability not only to transform into different cell types but also to mitigate inflammation. A phase 2 study, which analyzes mostly safety, was recently completed in patients with ARDS and showed positive findings.[11] Further studies are underway, and the entire pulmonary community is holding its breath to see whether the first treatment to improve outcomes for ARDS patients is on its way.

John B. West thought a lot about the issues of oxygenation and ventilation throughout his long career, from the quiet of his laboratory to the windy heights of cold Mount Everest. For the first five decades of his career, Dr. West studied the same class of animal—mammals. But then he turned his attention to a completely different species—birds—and brought awareness to important aspects of breathing.

Today, some ten thousand species of birds exist on Earth, about twice the number of mammal species. They colonize many different habitats and are able to maintain incredibly high workloads, or metabolic rates. One species that stands out is the hummingbird, which, with a wing beat frequency of up to 70 beats per second and a heart that can go to over 1,200 beats per minute, has a metabolic rate thirty times higher than that of humans. Another remarkable bird is the bar-headed goose, which is able to fly up to thirty thousand feet. These are feats of physiology that humans could not think of matching, and Dr. West believes it is their bird lungs, radically different from human lungs, that allow them to sustain these very high workloads.[12]

Dr. West's interest in birds was piqued in 1960, when he spent six months with a team of researchers on Mount Everest. The project was dubbed the Silver Hut Expedition for the tin house in which they lived, on the Mingbo Glacier, at nineteen thousand feet (the nearby peak of Mount Everest is at twenty-nine thousand feet). From this perch, West and the other scientists investigated the effects of high altitude on the human body. After some time in this environment, West had become frightfully tired and thin from the stress of altitude. One morning, as he struggled to get going, he looked out of the window of the Silver Hut, drawn by a quacking noise. Way above his head, at about twenty-one thousand feet, was a gaggle of twelve rather ordinary-looking tan geese flying effortlessly in skies normally reserved for jet airplanes. How could

West explain the difference between his own extreme fatigue and the bird's easy flight?

The answer to his question lay in the design of their lungs. Despite all the obvious differences between birds and humans, one of the most important distinctions is not immediately apparent, though it is the likely key to birds' success colonizing so many habitats: their lungs have separated out the jobs of oxygenation and ventilation. Our lungs are simple in that they have combined these jobs into a single unit. They expand and contract to provide movement of air, or ventilation, much like a fireplace bellows. The same areas that expand and contract to provide ventilation also house the gas exchange areas that allow oxygen to move into the blood and carbon dioxide to be released.

But as West's observations led him to point out, if an engineer was designing a breathing machine, the functions of gas exchange and air movement would be separated. Birds have such a system. With each breath they take, air moves into air sacs, which are large, easily distensible organs in which no gas exchange takes place. This air then gets shunted to a separate area for gas exchange, termed air capillaries. There, because there is no need for the gas exchange units to bend when a breath occurs, the distance between the air and the blood vessels is incredibly thin, much less than the one third of a micron in mammals, making the exchange even easier. A final difference is that air movement in the bird lung goes around in a circle, much as blood does, so birds get fresh air with both inspiration and expiration. We, in contrast, are limited to getting fresh air only with inspiration.

With all of these differences, Dr. West argued that a bird's breathing system is more efficient than a human's. At the same time, our form follows our function. For most of our needs, the lungs we have do an excellent job. It's only when we follow the bar-headed goose up Mount Everest with no supplemental oxygen, as some have attempted, that a set of bird lungs would come in handy. Biology has set limits, which some people try to ignore, at times to their own detriment.

The design structure of our lungs ties into our method of locomotion and basic needs of survival, which are much different from those of

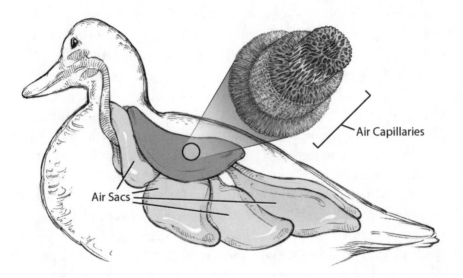

Anatomy of a bird's lung with air sacs for ventilation and air capillaries for gas exchange.

birds. But for all the variance between our classes, the standard blood levels of oxygen attained in a mammal and a bird are, surprisingly, exactly the same—about 95 mmHg in both species. Both bird and mammal systems seem to be tied into this optimal level of oxygen, not too low but not too high.

We know the problem at low levels of oxygen, but at higher levels, above 100 mmHg, oxygen may become toxic by grabbing electrons we don't want it to grab, in a process similar to the oxidation that produces rust on a car. The laws and limits of chemistry are at work in all of our biological systems, including the gross anatomical structure of the lung.

It's not just the laws of chemistry that govern our physiology. Other forces of nature are at work as well. As described in the prologue, our lungs resemble a tree branching up into leaves or down into roots. The lungs could also be compared to the tributaries of a riverbed merging to a main waterway. The neurons of the brain, spreading out into tendrils from the main axon, follow a similar configuration. The human body itself is another example, dividing from a main trunk into limbs, which then divide into fingers and toes.

This branching configuration is so familiar because it appears to follow a law of physics first described in 1996 by Duke University physicist Adrian Bejan. Termed the *constructal law*, it states that, "for a finite-size system to persist in time, it must evolve in such a way that it provides easier access to the imposed currents that flow through it."[13] The best structure for this happens to be how our lungs are designed, with many small branches connected to one big branch. The lungs are tied into the universe of physics like no other organ, perfectly using the space allotted to maximize flow. And optimizing flow, and movement, is clearly one of the purposes of life from a biological perspective.

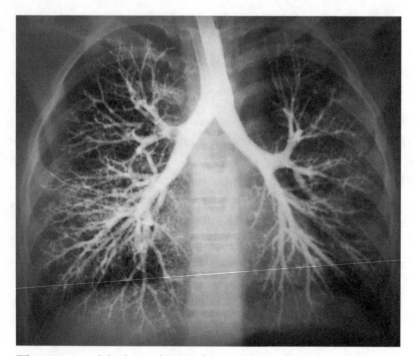

The airways of the lung, designed to maximize flow.

Chapter 3

An Infant's Drive to Breathe

The lungs not only facilitated the beginning of terrestrial life on this planet, they also facilitate the beginning of our individual lives. During the final trimester of pregnancy, the lungs are the only organ in the fetus that is not working. The heart is beating away at a fiery 160 beats per minute, the kidneys are making urine, with the baby peeing right into the amniotic fluid (which is then swallowed again by the baby in a repeating cycle). The brain and muscles are awake with kicks and backflips and rolls. But the lungs remain completely silent and nonfunctioning.

This all changes, and abruptly, when the baby emerges from the womb. The lungs must turn on in an instant in order to begin their job of oxygen extraction and carbon dioxide release. To measure the success of this change, all hospitals throughout the world use what is called the APGAR score, named after the distinguished professor of surgery from Columbia University, Virginia Apgar. The first woman to attain full professorial status at Columbia's medical school, Dr. Apgar devised this beautifully simple and elegant way to assess the health of a newborn in 1953.

At one and five minutes after birth, the Appearance, Pulse, Grimace, Activity, and Respiration are assessed, and a score of 0, 1, or 2 is recorded for each, for a maximum of 10. The majority of babies easily achieve a score of 8 or 9. The purpose of the APGAR score is to identify babies at immediate risk and to take proactive measures to improve whatever deficiency is present. This could mean just agitating the baby until he or she wakes up, or it could mean giving more oxygen or inserting a breathing tube into the lungs. Sometimes in medicine inaction is preferable, along the lines of "First do no harm." A low APGAR score is not one of these times. A newborn scored at 6 or 7 will usually improve on her or his own. A score under 5 is panic time.

The APGAR score reflects, on the most basic level, the ability of the lungs, heart, brain, and muscles of the chest to appropriately make the lightning jump from living in fluid to living in air. But of these four systems, the lungs have by far the most work to do, because in utero they are like a soaked sponge, filled with the mother's amniotic fluid. The fetus's source of oxygen is the placenta, that radiant red jellyfish-like structure that is expelled after birth. The placenta neatly takes oxygen-rich blood from vessels imbedded in the uterus and channels it to the fetus through the umbilical vein.

Once in the umbilical vein, the blood travels through a series of open ducts, one through the liver and another through the right to the left side of the heart, to ensure that the dormant lungs are bypassed. The blood then goes out the left side of the heart to the aorta, where it feeds the organs. The tissues expel the oxygen-depleted blood back into veins, where it ultimately travels back into the mother's body through the umbilical artery.

The free oxygen ride must come to an end, and it does so dramatically at birth. In an instant, the ducts through the liver and heart close, shunting blood to the lungs to pick up oxygen. The brain must simultaneously start firing signals to the muscles of inspiration. The eyes must open and adjust to the harsh light of the world. Finally, the lungs, still filled with amniotic fluid, must inflate in an instant with the first breath of life. The alveoli pop open for the first time and, with that first deep breath, suck the fluid up and immediately begin extracting oxygen from the atmosphere. The lungs change from being water-filled to being air-filled, from being dormant to extracting oxygen, all in the first few seconds of life.

Unfortunately, for some babies this leap from in utero to living in the atmosphere is not without significant complications. I experienced this firsthand one day. On a brutally hot day in late spring, I drove frantically to the hospital in dense Philadelphia traffic with my very pregnant wife. Adding to the discomfort, my wife was intermittently squeezing the blood out of my arm in retaliation for the contractions in her belly.

We drove right up to the hospital and gave the attendant my key. A man instantly emerged with a wheelchair, and we were whisked up to the preadmission area for pregnant patients. A nurse in bright-green scrubs immediately and unceremoniously put a glove on her hand and inserted it into my wife. "You're almost completed dilated," the woman said. "We need to get you into the delivery room. Now!"

Our hearts started racing as we instinctively clutched each other's hand. The nurse left, but not for long, and when she came back, she was accompanied by a horde of other hospital workers. With mechanical efficiency, one of them jabbed my wife's arm for an IV line, another thrust a blood pressure cuff around her bicep, and a third strapped a monitor onto her belly to measure the baby's heart rate. Then she was quickly moved to the delivery room, up onto the bed, and into position.

"What about my epidural?" my wife asked, squeezing my arm again as another harsh contraction pulsed through her. The doctor came in, young and fresh-faced, in blue scrubs and blue hat. She nodded at us and then studied the baby's heart rate on the monitor. It had dipped down with the contraction, which was normal enough, but it was going too low, and staying too low for too long. After a long spell of slow, low-pitched, tortuous beeps, the pinging of the heart rate on the monitor resumed its brisk pace.

"Listen, there's no time for an epidural. You need to get this baby out. He's ready. Your body is ready. We need to do this."

"You're sure?" My wife looked anxiously around, stressed by the prospect of more pain.

"Yes, quite sure," the doctor responded evenly. "We need to get this little guy out. There's something irritating him in there. His heart rate is intermittently dropping too low. Way too low. He needs to come out now."

Wild thoughts entered my head. He was a few days early, and now his heart rate was sporadically bottoming out. Questions about whether this would affect his brain, and whether his lungs would be ready to wake up and answer the call of terrestrial life, entered my head.

For the next fifteen minutes, my wife's contractions came and went. With each one, the little guy's heart rate dipped too low and for too long. But it always came back up, granting us some feeling that everything was okay.

Finally, in response to one long, very painful contraction and a lot of pushing, the baby's head appeared in the canal, his hair all curly and slimy. "Okay, let's do it on the next one," the doctor said, now fully suited in blue paper scrubs and elbow-length white gloves, her energy raised to the next level.

The next contraction came, and through the searing pain and exhaustion, my wife screamed and pushed, completely absorbed in that private world of childbirth. But from her pain and monumental effort came a result—my son's head popped out. The excitement was tempered by the sounds of his heart monitor, which began bleating out the low drone of his heart rate crashing again. It dropped much lower than before, down to forty beats a minute. My wife stopped pushing, the contraction gone. Her face relaxed, and then her pelvis. The baby retreated to where he had come from, and the slow heart rate that should have started to recover by now didn't. And it was dropping lower and lower, now thirty, now twenty, with no signs of recovery. Then the pinging heartbeat dropped to its lowest and slowest pitch yet, a sickening sign of a life slipping away.

"Don't stop! Don't stop!" the doctor pleaded, grabbing my wife's hand in hers. "You need to get this baby out. Push! Push! Push!" I joined into the entreaties and started screaming, "Push! Push!" Confused, and shielded from the reality of her newborn's devastatingly low heart rate, my wife started pushing again. Once and nothing; then twice and nothing.

"One more big push!" I yelled. It was my turn now to squeeze her arm, and I squeezed it hard. Finally, with a huge effort and a high-pitched scream, my wife pushed with all of her might, and the little guy came squirting out in rush of liquid and slime. He was beautiful, but he wasn't moving at all. His head and body were completely flaccid, his eyes shut, and his skin a sickening pale blue.

It was clear now what had caused his heart rate to dip low: around his neck the umbilical cord was wrapped tight in a single, well-defined knot. It all made perfect sense. As he got farther down the birth canal, the umbilical cord, constrained by its tether to the placenta, had wrapped itself tighter and tighter, like a perfectly constructed noose.

The nurse quickly cut the cord, and she and the and doctor brushed past me and put the baby on the newborn bed, a bright warming light shining down.

"Somebody page pediatrics. *Stat!*" the doctor yelled. "His APGAR is four."

She then turned the warming lamp up to high and shook our child's chest. He did nothing, remaining flaccid and blue. The doctor grabbed the oxygen mask and strapped it onto his face, but still nothing happened. Ten seconds passed, then twenty, then thirty, without the faintest hint of a limb stirring.

A nurse hurriedly got an intubation tray ready to insert a tube into my son's mouth and hook him up to a ventilator. If he couldn't breathe, the ventilator would have to do it for him. I took a look at the instruments the staff was about to employ. The intubation blade, about six inches long and shiny silver, would be used to pull open his mouth to get a good look at the airway opening. The tracheal tube was a simple piece of plastic with a balloon on the end, ready to be thrust down to deliver the breaths of life. There was no question about him needing the tube. We were just waiting for the pediatricians now.

Another nurse came over with a breathing bag to deliver rescue breaths until they got the tube in. Before strapping on the mask, she gave my son one final shake, and through a miracle, she connected with what was likely the single neuron in his brain that was still firing. He shook his head, took a huge breath, instantly turned a bright red, and let out a huge scream, affirming his secure place in the world.

* * *

On the spectrum of childbirth complications, my son's issue was serious, but an umbilical cord around the neck is not uncommon. In the 1950s and '60s, the problem facing pediatricians in the United States was significantly worse, with ten thousand newborns a year dying of a mysterious lung disease, not to mention the other thousands worldwide. Most who succumbed didn't live past a week. In the United States, another fifteen thousand who were affected by this strange inflammatory condition were left with suboptimal lungs when they recovered.

Typically, these little ones were born early, sometimes by a few weeks, sometimes by a few months, and never got a chance in life. Their deliveries were generally uncomplicated, but within a few minutes of birth their breathing would become labored and noisy. High-pitched grunting would come out of their lungs with exhalation, and their nostrils would flare out and in as they struggled to get enough air into their lungs. Their chest walls would pump up and down, their breath rapid and shallow. Their skin, initially a healthy pink from their mother's oxygen supply, would turn a grayish blue, the tips of their fingers forbiddingly dark. Other complications followed—bleeding into the brain, kidneys shutting down, and seizures.

From the delivery room, the babies were attended by pediatricians who desperately tried to keep them alive. But there wasn't much the staff could do, since not much was known about how to treat them at the time, and no medicines existed to cure whatever was happening. And so these babies, often very small but with normal hearts and brains and kidneys and livers, had their lungs collapse for no apparent reason. Many died.

The most famous of these breathing-challenged babies was Patrick Kennedy. Born five and a half weeks early on August 7, 1963, on Cape Cod, he began having breathing difficulties immediately after birth. Transferred to an intensive care unit (ICU) in Boston, he continued to decline, his organs failing. His body finally gave out and he passed away two days later. If there was nothing very remarkable about the baby's illness, there was something unique about his parents. His father was John F. Kennedy, thirty-fifth president of the United States, and his mother was Jacqueline Bouvier Kennedy, the First Lady.

The nation mourned with them that August, but that was all any-body could do, because nobody had a clue as to what was causing these tragedies.

Mary Ellen Avery, who eventually helped solve the mystery of neonatal respiratory distress syndrome, came from a simple background—her mother was a school principal, and her father, despite being blind, started a successful cotton products business during the Depression in the 1930s. The lesson he taught his children was obvious: problems were meant to be solved.

Mary Ellen started kindergarten early, and then skipped sixth grade. By seventh grade, she was telling everybody she wanted to be a doctor. This desire was no doubt due to the influence of the seventh grader's neighbor and mentor, Emily Bacon, a professor of pediatrics at the Woman's Medical College of Pennsylvania, in Philadelphia. Dr. Bacon would take Mary Ellen with her to the hospital some mornings and show her newborns in the nursery. There, one day, Mary Ellen saw an infant grunting and wheezing and turning blue, her first exposure to premature respiratory distress syndrome. If this disease could be cured, she thought, the additional years of life added would quite simply be a lifetime.[1]

Mary Ellen attended Wheaton College in rural Norton, Massa-chusetts, where she continued her upward trajectory, graduating *summa cum laude* in chemistry in 1948. Determined to get the best medical education possible, she applied only to Harvard and Johns Hopkins. She didn't know then that Harvard didn't accept women, and she never heard from them. But Johns Hopkins University School of Medicine had been founded in 1893 with money from several wealthy female benefactors, who had insisted that educating female physicians be an equal part of the institution's mission. They accepted eighty-six men and four women in Mary Ellen's year.[2]

Despite the challenges and resistance from some chauvinist profes-sors, Mary Ellen graduated and stayed on afterward for an internship and residency in pediatrics. A month into the internship, in a screening test, she was diagnosed with tuberculosis and was packed off to a sanatorium in upstate New York, where she was instructed to lie down for most of the day while the antibiotics did their job. Once cured, she returned to finish her training at Hopkins in 1954. The hours were long—shifts of thirty-six hours were the norm then—but it was an exciting time to be in medicine. A year earlier, in 1953, James Watson and Francis Crick had written a paper on the structure of DNA, our genetic material. Also around this time, cardiac catheterization started, and accurate diagnosis of heart disease became a reality. The number of available antibiotics expanded to five, then ten, then twenty. Huge medical breakthroughs seemed to be coming once a month.

At the end of her three years of clinical training, Mary Ellen was still deeply bothered by babies dying of lung failure, and held to the dictum of the Italian Renaissance scientist and philosopher Galileo Galilei: "I would rather discover a single fact, even a small one, than debate the great issues at length without discovering anything at all."[3] The single issue she wanted to investigate was why these premature babies' lungs didn't work at birth, and what was different between a thirty-two-week-old newborn's lungs and a forty-week-old newborn's lungs. She decided to work with Jere Mead, who was doing seminal work in pulmonary physiology at the Harvard School of Public Health in Boston.

The disease that we now call respiratory distress syndrome of the newborn had many different names in the 1950s, including congenital aspiration pneumonia, asphyxial membrane disease, desquamative anaerosis, congenital alveolar dysplasia, vernix membrane disease, hyaline membrane disease, and hyaline atelectasis. Most doctors today can't tell you what most of those words even mean. But the esoteric names sprang from the many theories of the syndrome's cause, masking the unknown in obscure language. Some believed the infants were breathing fluid into the lungs as they passed through the birth canal. Others hypothesized a heart defect, which was causing fluid to back up into the lungs. Another theory proposed that pulmonary circulation was the source of the problem. Unsurprisingly, clinical trials for potential medicines in human subjects all came back negative.

Despite how far the entire field was from solving this problem, a few things were known. Autopsies noted that the alveoli, those small grapelike clusters where gas exchange takes place, were plugged up with dead inflammatory cells and protein waste, which were named hyaline membranes. This material had a slightly transparent, glassy look. The term *hyaline membrane* came from the Greek word *hyalos*, meaning "glass or transparent stone such as crystal." Most scientists focused their research on this phenomenon.

Mary Ellen, now Dr. Avery, deliberately did not focus on hyaline membranes, or any other existing theory, freeing herself from all preconceptions and throwing herself into understanding the basic physiology of the lung. Her approach, like that of most of successful scientists, was to explore the mechanisms underlying a given process and not just to observe the output. She focused on the basic questions of what allowed the lung to expand and contract, over and over and over again, without being ripped apart or collapsing in on itself, on what gave this wonderful organ its resiliency and strength to breathe 20,160 times per day, moving some ten thousand liters of air, while an additional five liters of blood makes its way through the blood vessels of the lungs every minute. The heart is made of compact, strong muscle. The liver is a dense structure of channels and filters. The lung, by contrast, is mostly air. Under a microscope, it has a thin, lacy structure, delicate in appearance. Where its resiliency and strength came from was a mystery.

Dr. Avery studied the respiratory physiology of different animals from birth to a few weeks old, mapping their lung development and characteristics as they emerged into life. Away from the lab, she continued her clinical work at the Boston Lying-In Hospital, overseeing the care of the newborns. Obstetricians would hand the newborn babies to her, and she would start a stopwatch and write down the data as the baby inhaled for the first time, calculating an APGAR score and then taking blood samples. She ran from room to room, her mind on high alert for any clues about these babies' lungs.

When babies died from mysterious lung illness, Dr. Avery was at their autopsies, going over their pathology, holding on to the slides for the day when she could make more of a connection between them. One thing that caught her attention during these autopsies was how dense with tissue these little baby lungs were, completely airless, resembling the liver more than the lung. They had failed to inflate.

Dr. Avery visited the library at the Massachusetts Institute of Technology (MIT) on weekends, seeking literature from fields outside of medicine, hunting for new ideas from the minds of chemists and mathematicians. On one of these visits she discovered a book by C. V. Boys entitled *Soap Bubbles: Their Colours and Forces Which Mould Them*.

First published in 1912 for English schoolboys, this slim volume was a primer on the physical properties that govern soap bubbles, filled with simple experiments that document the physical properties of liquids and their interaction with air, explaining how soap bubbles are able to stay intact, miraculously floating through the air. Dr. Avery saw a connection between soap bubbles and the alveoli in our lungs. Circular in shape, and needing to stay open to continue gas exchange, alveoli are governed by the same physical laws as those governing soap bubbles.

The key to soap bubbles staying spherical and not collapsing in on themselves lies in their surface tension. Any spherical structure, like a soap bubble or an alveolus in the lung, is bound by a simple law of physics. Formulated by French scientist Pierre-Simon Laplace and English mathematician Thomas Young in 1805, the law states that the pressure exerted on a circular structure is directly proportional to the surface tension in the sphere, and inversely proportional to the radius of the sphere. Extrapolated out, this means that larger bubbles are more stable and have less pressure on them than smaller bubbles, and they are more likely to stay intact. Similarly, a sphere with lower surface tension is more stable and is under less pressure than one with higher surface tension.

The radius of a sphere is simply the distance from the center of the sphere to any edge. Surface tension, however, is more complicated. At the interface between a liquid and a gas, the molecules in the liquid are more tightly bound together than in other areas of the liquid. For example, in a glass of water the water molecules at the surface are much more crowded together than the molecules in the middle of the glass, because there are no water molecules above them to exert a dispersing force. These tightly bunched water molecules at the surface cause tension, which produces the slight dip one can see at the top of a glass of water.

Different liquids have different tendencies to bunch together at the surface. Water has a relatively high surface tension, so molecules are bunched relatively tightly together at its surface. Consequently, water does not make a good bubble, and exists more easily in drops, like rain drops and drops of water in a sink. But if soap is added to water, the surface tension is dramatically lowered. The ends of soap molecules have different properties: one end attracts water (hydrophilic), and the other one repels water (hydrophobic). When placed in water, the hydrophobic ends of soap molecules push their way to the top, which causes the water molecules to separate from one another, lowering the tension and energy between them. This allows a spherical structure like a soap bubble to stay intact, until it dries out and bursts.

At the same time Dr. Avery was learning about bubbles and surface tension, a number of committed scientists, employed by the federal government at the height of the Cold War, were investigating properties of the lung in reaction to chemical warfare. The lungs are a typical entry point for poisonous gases, and understanding the effects of toxins on the lung and how to combat them was a priority. One of these researchers, Dr. John Clements, at the Army base in Bethesda, Maryland, undertook a series of experiments in the mid 1950s to quantitatively measure surface tension in the lung, which demonstrated that lung tissue had very low surface tension compared to other tissues. He then did something simple, which nobody had ever done: he measured pressure across extracted lung tissue with expansion and contraction. As mentioned, the pressure on a sphere like a soap bubble or a lung alveolus is proportional to its surface tension divided by its radius, and lower pressures will mean the bubble will have a greater chance of not collapsing in on itself. Remarkably, the pressure decreased significantly with lung contraction (as the alveoli in the lung were getting small with contraction, pressure should have increased as the radius decreased), and increased significantly with expansion (as the alveoli got bigger, pressure should have decreased as the radius increased). To explain this, Dr. Clements correctly postulated that something must be overcoming the effect of size on pressure, and the only variable left in Laplace's equation was surface tension.[4]

Taking his hypothesis further, Dr. Clements imagined that something within the lung must be lowering surface tension so dramatically as to overcome the effect of size on pressure. He correctly postulated it was a soap-like foam, which exerted a dispersal effect as its molecules became more concentrated and the area became smaller, and lost this effect when the lung expanded and pulled the soap like foam molecules apart. The effect of this soap-like foam lowering surface tension would be more important than lung size in calculating pressure if it was a powerful substance (which it was, and is). John Clements later named this substance *surfactant*, from its effects on the surface tension.

The definitive discovery and demonstration of the existence of surfactant was a major breakthrough in the understanding of lung physiology, finally explaining the mechanism by which the lung seamlessly expands and contracts, thousands of times a day, without breaking apart with inspiration or collapsing with exhalation. While the heart has dense striated muscle, and the brain its conglomerated networks of communicating neurons, the lung is a thin, graceful structure of interconnecting fibrous tissue that is beautifully held together with a foamy substance that lubricates its functions in a quiet and effortless manner. It is an organ of elegance, not brute strength.

John Clements's paper did not get accepted into the high-powered journal *Nature*, but appeared instead in a low-level publication, where it was not widely recognized as the landmark study it would become.[5] It did, however, reach Dr. Avery, and in 1956 she drove to Bethesda to meet Dr. Clements in person. He knew nothing about neonatal respiratory distress, and she knew nothing about how to properly measure surface tension. He taught her everything he knew about lung physiology, as

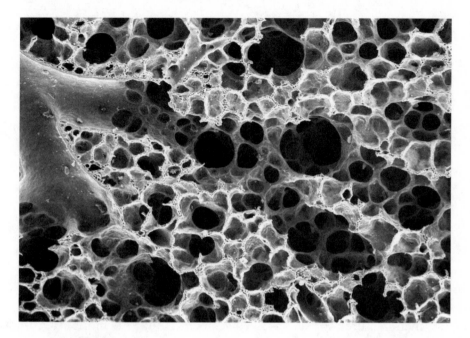

The lungs in cross section, with a conducting airway surrounded
by many alveoli.

well as how to build an instrument so she could take her own pressure
and surface tension measurements. Dr. Avery quickly came to believe
that the afflicted newborns weren't diseased because of the *presence* of
something, that something being hyaline membranes, but because of
the *absence* of something.[6] That something, she believed, was surfactant.

  She went back to her lab and built her own balance to measure
surface tension, and then she discerned that the lungs of babies who had
died from respiratory distress syndrome had very high surface tension.
By comparison, the lungs of normal infants had a much lower surface
tension. This was the breakthrough she had been looking for since her
time as a child visiting the hospital with Dr. Bacon, and the breakthrough
humanity had been waiting for since the first premature baby had been
born and died a perplexing death.

Dr. Avery published her findings in 1959 in the *American Journal of Diseases of Children*. Entitled "Surface Properties in Relation to Atelectasis and Hyaline Membrane Disease," the paper broke the field of neonatal respiratory distress syndrome wide open.[7] The key to the disease had been pinpointed. The immature lungs were not making surfactant, the surface tension in the alveoli was way too high, and the alveoli were crashing closed. Hyaline membranes were formed as a byproduct of the inflammation and destruction. Some babies lived long enough for surfactant production to kick in and open up their alveoli, but many did not.

Funding poured in from the National Institutes of Health, and over the ensuing decades, researchers at several different institutions made significant progress toward a cure. Doctors used ventilators to stent the lungs and alveoli open, and steroids were shown to speed up surfactant production in premature babies. Later, an artificial surfactant was manufactured to serve as a replacement. Today the mortality from respiratory distress syndrome is 5 percent of what it was before Dr. Avery's brilliant insight.

Mary Ellen Avery went on to accomplish other great things in her life. She helped found the field of specialized care for the newborn, known as neonatology, and her textbook, *Avery's Diseases of the Newborn*, has been the standard in its field for decades. She became a full professor of pediatrics, and the first female chief of a clinical department at Harvard Medical School. Her guidance produced tens, if not hundreds, of leaders in pediatrics across the country.

As for my son, all has gone well since his delayed first breath of life. He made the difficult transition from living in water to living in air. That day, my son taught me that breathing can be difficult. We take it for granted, but it is a complex process involving the coordination of multiple organs, with the lungs in the center. And a lung is not just a simple pump, pushing gas around. As Dr. Avery began to teach us, it is an organ alive with immunology and chemistry, one that does an extraordinary amount of work under extreme stress from the moment we enter this world.

Chapter 4

The Extraordinary
Healing Power of the Breath

There is a problem with our health today that relates to how we practice medicine, but even more importantly to how we take care of ourselves and each other. This is reflected in the most basic measure of our nation's health, our life expectancy, which flatlined between 2010 and 2018, after decades of steady improvement. The 2010s, despite all of our advances in medicine and pharmaceuticals, will be a lost decade in terms of the overall health of the nation. The CDC blames two preventable conditions: an increase in drug-overdose deaths, from less than ten thousand in 1990 to over seventy thousand in 2017, and an increased suicide rate, from ten per hundred thousand in 1999 to fourteen per hundred thousand in 2017.[1]

The incidence of depression has also recently increased, from a rate of 6.6 percent in 2005 to 7.3 percent in 2015.[2] The rise among twelve- to-seventeen-year-olds is even more striking, from 8.7 percent in 2005 to 12.7 percent in 2015. In 2017, 3.2 million adolescents had at least one major depressive episode, which represents 13.3 percent of this population.[3] A myriad of related conditions, such as chronic anxiety, panic attacks, chronic pain, bipolar disorder, substance abuse, and ADHD are also plaguing adolescents. Our young ones and teenagers, who are exposed to so much at such a young age, are experiencing the greatest increases in these disorders.

We see this unfortunate trend in hospitals, where medical wards are filled with patients with preventable conditions. A 2020 study out of Johns Hopkins School of Medicine showed a total of 25 percent of medical ICU admissions were related directly to acute substance abuse issues and overdoses, accounting for 23 percent of total costs.[4] This study doesn't take into account admissions due to chronic medical conditions related to substance abuse, such as lung cancer from tobacco use or liver cirrhosis from alcohol consumption, which in certain populations make up 44 percent of ICU admissions.[5] Herbert Benson, professor of medicine at Harvard Medical School, has reflected broadly on this issue: "More than 60 percent of visits to physicians in the United States are due to stress-related problems, most of which are poorly treated by drugs, surgery, or other medical procedures."[6]

As noted by Dr. Benson, medicine has failed to respond adequately to the growing crisis. Increasingly, the practice of medicine has become more procedural, focused on genetics, the use of technology, and the coming explosion of artificial intelligence. In the most egregious examples, families have fought back, as when Ernest Quintana's family complained when he was admitted to the hospital for respiratory failure from COPD and was told by a robot consultation that he had no lung left to work with, and that he should focus on palliative care.[7]

Combating the disconnect between the practice of medicine and our declining mental health is going to take several approaches, but one tool that can help in the battle against this epidemic of preventable conditions in our society is an organ: the lungs. We know this because the lungs have been doing this healing work for thousands of years, and some people have updated this knowledge and applied it to our world today.

The healing power of the breath was recognized as far back as 7000 BCE, in the Zoroastrian religion of Persia, now Iran, where breathing exercises were routinely practiced. This tradition was carried to the West, where both the ancient Greeks and Romans regularly engaged in breathing exercises and reflection. Meditation and chanting have remained an integral part of Judaism, Christianity, and Islam. But while Western religions talk and write about the breath—the Holy Spirit, and *ruach*—Eastern religions have made a strict focus on the breath being a part of spiritual enlightenment.

With three hundred million practitioners, Buddhism is one of the most widely observed religions in the world. For Buddhists, including those who follow the offshoots Zen and Tibetan Buddhism, breathing exercises are one of the foremost ways to practice their faith, the core from which all other habits, and ultimately enlightenment, spring. Buddhists believe you start with the breath, and the body and mind follow.[8]

Buddhism began with Siddhartha Gautama, a monk commonly known as the Buddha or Enlightened One, who lived on the Indian subcontinent sometime in the fifth century BCE. At age thirty-five, in a reaction to stress in his life, he set out to find enlightenment, learning from the local yogis of his time. After his training, he began teaching a unique path to spiritual insight. His ultimate goal was to achieve inner peace and understanding, a state he called nirvana.

The teachings of Buddhism are summarized in four noble truths, all of which need to be accepted and followed in order for progress to be made. The first is an acknowledgment that pain and suffering are a regular part of life. It is normal to feel anger, disappointment, loneliness, and frustration. Second, we should understand that suffering comes from desire, from being disappointed in ourselves and others by creating expectations, from constant wanting and cravings. Destructive behavior originates from letting the emotions of life take over our thoughts in a damaging manner. The third truth is that if we are able to free ourselves from these detrimental thoughts, we will obtain wisdom and insight. This end goal is also called nirvana, which literally means *blown out* or *extinguished*—the extinguishing of our attachments.

The first three noble truths identify the problem (suffering), illuminate the cause (attachments), and describe what happens when the problem is solved (wisdom and nirvana). The fourth noble truth lays out the path from attachments to wisdom. This is the Eightfold path, or eight teachings to guide one's journey. The first two teachings ask one to follow a path of wisdom by committing to the Buddhist path and pledging oneself to a moral life. The next three teachings show that ethics in behavior must be followed in speech by avoiding gossip and speaking truthfully; in action by refraining from killing, stealing, and overindulging in sensual pleasure; and in livelihood by not killing animals or trading in weapons or intoxicants. The last three teachings are dedicated to meditation. An effort must be made toward positive thinking, developing awareness of one's body and feelings, and finally, improving concentration through enhanced mental focus. *Commit, act, then practice and develop.* This is the eightfold path.

The Buddha taught that the primary way to get to a calm and enlightened state was by utilizing one's lungs. His teachings are captured in the text entitled *Anapanasati Sutta*. *Anapanasati* means "mindfulness (*sati*) of breathing (*anapana*)" and is the chief method by which insight is attained. According to the text, one should find a quiet place to sit alone, perhaps outside underneath a tree, and then begin to notice the breath, to concentrate on the inhales and exhales. If the breath is short, notice that it is short, and if long, notice that it is long. Later, more advanced breathing exercises can be practiced. Through this focus on the breath, one is able to push out other thoughts and begin to give attention to one's mental capabilities, to release attachments and expectations in the quest for understanding.

Hinduism, another Indian religion founded in the fifth century BCE, also has at its core attending to the breath as a method to achieve enlightenment, mostly through the practice of yoga. One of the main yogic breathing practices is *pranayama*, a combination of the two Sanskrit words *prana*, meaning life force or vital energy, and *yama*, which means to extend or draw out. It's a method aimed at achieving inner peace and control. The different methods of practicing *pranayama* all involve a focus on the breath, which is the source of our life force, our *prana*.

In the West in recent years, the teachings of Buddhism and Hinduism have been translated into the "mindfulness movement." One of the first to bring the idea of mindfulness to the West was Thich Nhat Hahn, a Vietnamese Buddhist monk who lived in exile in France for many years. There, he counseled that "whenever your mind becomes scattered, use your breath as means to take hold of your mind again."[9]

Despite this sound advice, the mindfulness movement has not been without controversy and criticism, with some calling it too divorced from its Buddhist roots, too much of a quick fix that has been tainted with a corporate edge. Some of this may be true, but the benefits outlined below show that a concentrated and dedicated focus on the breath has clear benefits.

If one tells a primary care physician today that one has depression, chances are a prescription for an antidepressant medication will be forthcoming—with little discussion about the cause of the hopeless feelings. TV producer Amy Weintraub knows the limits of this approach. For many years, she was stuck in a cycle of self-doubt, lack of pleasure, and decreasing energy. Exercise and coffee didn't help, nor did a stable relationship or medication. Even with a successful career, Amy was entering into middle age without a sense of direction or purpose. She wasn't experiencing what Virginia Wolf called "wave after wave of agony," but rather what Emily Dickinson called "an element of blank," experiencing a fog over her thoughts, a layer of cotton between her brain and her skull.[10]

Autumn always seemed to be Amy's low point, as the New England sky turned dark, leaves disappeared, and the cold and rain forced everybody inside. She remembers the fall of 1985 as a particularly bad time, because a hurricane threatened her and her partner's house in Newport, Rhode Island, and despite the impending storm, she couldn't find any energy.

A few weeks later she sat on her therapist's couch in Providence and explained her extreme inability to feel pleasure. Her therapist observed that perhaps Amy would always experience "empty pockets." Driving home afterward, Amy felt she had no choice but to stay on medication, try to get out of bed in the morning, and otherwise do the best she could.

This continued until one day, while collecting her absent neighbor's mail, Amy noticed a catalog from the Kripalu Center, in Stockbridge, Massachusetts. The center offered yoga classes and retreats, and Amy signed up for one, despite having low expectations. Her expectations couldn't have been more wrong: the three days in Stockbridge ended up being a revelation of sorts, an awakening of her body and mind that would lead her on a path to recovery.

The first thing the yoga instructor at the Kripalu Center asked Amy and her class to do was stand up straight and tall on their mats, shoulders back, head high. Then she asked them to put their hands in prayer position in front of their hearts with elbows out to the side. The instructor then told them, "Take a deep breath in and fill your heart with light. Hold the breath and feel the light as healing energy expands through your chest and through your whole body. Exhale and open your palms to receive. Stay empty. God loves your empty hands."[11]

Over the ensuing hour, and with each pose and breathing exercise, Amy felt the slow awakening of her body and mind. Her shoulders relaxed, her heart expanded with positive feelings, and beams of light seemed to shoot through the tips of her fingers and toes. Insights rushed into her now calm mind, which before had been blank, suggesting the emptiness she had been experiencing was really an opportunity—in this case, to experience divine parts of life and to know that she was indeed worthy of affection and happiness.

After several days of participating in yoga classes, workshops, and evening music and chanting sessions, Amy returned to Newport in a rejuvenated state of mind, and continued doing the work of postures (asanas), as well as breathing exercises (pranayamas), with video tapes. Some mornings it was still difficult to get out of bed, but they were fewer, and on the bad mornings she did her breathing exercises in bed. Easy breath in, hold it, and easy breath out. A few months later, another big leap forward occurred while Amy was in the car, listening to a guided imagery tape. At the end of the tape, she was asked to "name" herself. The name that came to her was *abundance*. Her pockets were full now.

After recovering from her own depression, Amy gave up her successful career as a television producer and moved on to the higher calling of yoga teacher. For years she led 6:00 a.m. yoga sessions, encouraging people to just "get on the mat," telling her students that yoga class is a nonjudgmental space, especially when it comes to internal voices of criticism. Accept life's limitations and focus on what's right with yourself, not what's wrong. Acknowledge your feelings as valid, but don't be controlled by them. Avoid too much introspection—focus on the breath and the body, and the mind will follow.

In her book *Yoga for Depression*, Amy tells stories of transformation that astonish those in traditional medicine. People come to her class the first time with shoulders hunched over, shallow breathing from their upper body, and eyes averted. In a few months, many are standing upright, breathing deeply from their belly, looking people in the eye, and smiling not with a mask on, but with the true smile of somebody at peace.

Foremost among those who have attempted to bring breathing and relaxation into the modern Western consciousness is Jon Kabat-Zinn. He sees mindfulness as the antidote to the information overload we all experience from cell phones, the Internet, and TV, as a way back to our families and loved ones.[12]

One day in 1972, while finishing his PhD in molecular biology at MIT, Kabat-Zinn saw a sign on campus for a talk given by a Buddhist monk. He attended the lecture, was transfixed by the monk's ideas, and set off on a path to study meditation and mindfulness, bringing along his scientific background as he learned about this ancient discipline. Several years later he set up a center for transformational medicine at the University of Massachusetts Medical School and began to help some of the many people who had slipped through the cracks of traditional medical care. He told his patients to accept what he calls "the full catastrophe of life" as the place from which to move forward. Patients with all types of illnesses, from chronic pain to anxiety to cancer to heart disease, flocked to the center. In his practice, with its focus on the breath and quiet meditation, Kabat-Zinn has seen radical transformations in people's lives that regular doctors can only wonder at.

Besides helping patients deal with illness, he also encouraged a change in how medical students are trained. He recognized that patients wanted doctors who not only could make diagnoses, but also could empathize with people. Teaching mindfulness and awareness to medical students was a big part of this. At my own medical school, I once attended a session in which patients expressed what they liked about their doctors. One patient said the most important thing to him was for his doctor to take a seat in front of him, look him in the eye, ask him simply, "How are things going?," and then pause and wait for an answer. The doctor wasn't typing on a computer, looking at her phone, or answering a page. According to one study, the average time it takes a physician to interrupt a patient is eleven seconds.[13] When a doctor can give patients just two minutes, it can make a big difference. This was the most important lecture I went to all year, and it deeply affected how I practice medicine.

Listening without judgement, being there in the moment, is at the core of the mindfulness movement, and it can be practiced in different ways. Some people enjoy breath exercises in the context of movement, as with yoga. Others may lie on their back and then go through a "body scan," consciously cycling through the body as a whole and then each body part, paying attention to the feedback from that part—its temperature, its texture, how it interfaces with the air around it. Another powerful technique is to pick a single object, like a raisin or a leaf, and observe it as if one had never seen it before—how it looks, feels, smells, tastes. It is not about clearing the mind, but rather observing in the present, having these exercises spill over into life outside of meditation so that one's awareness and appreciation open up.

The implications of cultivating mindfulness in medicine go beyond improved communication with patients. Much of medicine involves simple but careful observation, from doing a thorough physical exam to interpreting a chest X-ray or an MRI. Focused surveillance is a lost art, with the crush of data from technology and laboratory studies consuming much of doctors' time, let alone all the time spent tending to the digital medical record. One 2015 study at Stanford University demonstrated that 63 percent of diagnostic errors occurred because no physical exam was ever done, 14 percent of errors resulted from correct exam findings that were misinterpreted, and 11 percent of errors occurred because the physical sign was missed or not sought.[14] Mindfulness, with focused observation, could help guide us back to the bedside.

We now know that, physiologically, something important is happening during breathing exercises. The autonomic nervous system is deeply involved, which is the division of our nervous system that deals with everyday functions like breathing, heart rate, and workings of the gastrointestinal tract. The two main—and opposing—branches of this system are the sympathetic nervous system and parasympathetic pathways. The former gets turned on when one is scared, or under threat, and the hormone epinephrine (adrenaline) pours out from the adrenal gland, causing the heart rate to increase, the eyes to dilate, and sweat to rush out. This is the so-called fight-or-flight mechanism.

The parasympathetic system produces the opposite effect on the very same organs with the hormone acetylcholine—it calms heart rate and breathing rate, opens up blood vessels to the stomach, and provides a sense of well-being, appropriately termed *rest and digest*. Deep breathing is a potent inducer of the parasympathetic system. The release of acetylcholine not only calms our organs, it also stimulates the release of serotonin, dopamine, and prolactin, the feel-good hormones targeted by medicines like Prozac and Zoloft. But yoga and breathing exercises produce this effect naturally and without side effects.

The scientific literature showing how the release of hormones changes outcomes in many diseases, including those of the breath, is established and growing. Buteyko breathing, a method proposed in the 1950s by Ukrainian doctor Konstantin Buteyko, is aimed at getting control of hyperventilation by breathing through the nose, slowing down the breathing rate, and being aware of dysfunctional breathing. A randomized trial published in 2008 in *Respiratory Medicine* had asthma patients practice Buteyko breathing techniques, and at the end of six months there was an increase in those with well controlled asthma, from 40 percent of subjects to 79 percent.[15] Somewhat surprisingly, the subjects practicing breathing techniques were also able to significantly reduce their use of inhaled steroids. Another study, published in 2009 in the journal *Thorax*, showed breathing exercises lowered anxiety and depression scores for people with asthma.[16] While many asthmatics do need strong anti-inflammatory medicines to control their symptoms, it's clear that breathing exercises can play an important role in treatment.

The scientific evidence that these same breathing exercises improve mental health conditions and chronic pain is also increasing every year. A 2016 study involving ninety college students with depression and/or anxiety showed significant improvements in those who took either a yoga or mindfulness course, but not in the control group.[17] A 2014 study split sixty-four women with post-traumatic stress disorder into groups and sent them to either yoga or education about their condition, with yoga significantly improving symptoms compared to supportive health education.[18] A 2012 analysis involving 1007 subjects demonstrated that yoga improved pain symptoms significantly in patients with chronic disability.[19]

Breathing and stress-reduction exercises seem to help people suffering from mood and breathing disorders. But studies have also investigated how stress-reduction exercises affect the different inflammatory genes and proteins expressed in our bodies. A milestone study published in 2013 followed twenty-six subjects who had no prior experience with relaxation exercises as they underwent eight weeks of training. Blood samples were collected before and after their exercises and analyzed for the expression of different genes by measuring levels of ribonucleic acid, or RNA, the first structure made from DNA on the way to protein synthesis.[20] Compared to blood drawn prior to relaxation techniques, in blood drawn afterward there was a significant decline in RNA production associated with the inflammatory response, stress-related pathways, and even cell-death pathways (indicating these cells could potentially live longer). Genes that showed increased activity were those associated with improved energy metabolism, insulin secretion, and proteins that regulate genetic health and longevity.

A subsequent review paper pulled together thirty-four similar studies, recording inflammatory markers not only in healthy controls but also in subjects with leukemia, breast cancer, and dementia. Positive results in patients who practiced various breathing and meditative exercises were demonstrated across the board.[21]

The power of the breath has been used not just to heal, but to attain extraordinary feats that appear to defy laws of physiology. One of the most striking examples is the practice of g-tummo meditation by Tibetan monks. *Tummo* is the Tibetan word for inner fire, and the tummo meditation technique helps explain how the monks who practice it are able to sit outside overnight in the freezing cold of the Himalayas wrapped only in a thin sheet.

The techniques of "vase breathing" make this superhuman feat possible. Vase breathing requires one to sit quietly and focus on expanding the stomach as a breath comes in, imagining one is pouring water into a vase. Then, instead of exhaling all the way, one stops at about 90 percent of total exhalation, maintaining a round shape to the abdomen. When this practice is repeated in cycles, the increased ventilation that breathing with the stomach provides, along with the increased rate of breathing, has been shown to dramatically raise body temperature.

Wim Hof has recently taken the idea of g-tummo breathing to new heights, often quite literally. Born in 1959 in the Netherlands, he is nicknamed "The Iceman" for his Guinness World Record–breaking accomplishments in cold weather. He set the world record in 2011 for immersion in ice by staying submerged up to his neck for one hour, fifty-two minutes, and forty-two seconds. He has climbed Mount Everest to twenty-two thousand feet in only shorts and shoes, and he ran a full marathon in Finland, above the Arctic Circle, at -20 degrees Celsius, clad in a similar manner. With mastery of his breath and his meditative practices, Hof is redefining what is considered physiologically possible for a human.[22]

Take a moment to fall awake. Befriend yourself and your emotions. These and other mindfulness ideas can transform lives and be one approach to solving the daunting problems we face with increasing incidence of mental health disorders, deadly drug abuse, and depression. Changes in society are going to come at an increasingly rapid rate, and we need new tools and approaches to combat the stresses of the new reality. As Thich Nhat Hanh has said, "Feelings come and go like clouds in a windy sky. Conscious breathing is my anchor."[23] And more importantly, Buddhist Annabel Laity has counseled us: "Breathe! You are Alive!"[24]

PART II

THE PRESENT:
OUR LUNGS—AND US—
AGAINST THE WORLD

Chapter 5

A Window onto
the Immune System

With every breath, the wonder of gas exchange happens with apparent calm and ease. Underneath this calm, though, an incessant battle is occurring that we are only beginning to appreciate. The lungs have built up a complex immune system over millions of years to guard against the constant threat of invasion by viruses, bacteria, and parasites. Every minute, hundreds of different cells and cell types come and go in and out of our lungs as part of a finely coordinated effort with a simple goal—to kill invaders that mean us harm. Viewed in real time through a high-powered microscope, this battle can resemble a game of Pac-Man—a hungry inflammatory cell, such as the neutrophil, chasing a bacterium, each angling and weaving, with the determined neutrophil finally cornering, engulfing, and destroying the invader.[1]

Necessary to our existence, our immune system constantly keeps an array of invaders at bay. At the front line of this constant fight are the lungs, defending us from the influenza virus trying to move in every winter; from bacteria, like streptococcus and staphylococcus, attempting to infect us with pneumonia; and from tuberculosis kept at bay in the lungs of almost two billion people.

In recent decades, with improved sanitation and vaccinations, the threat from infectious invaders has been greatly reduced. Rates of measles, mumps, and Hepatitis A, along with many other infections, fell over 95 percent between 1960 and 1980.[2] On an evolutionary time scale this is extremely fast, and it appears our immune systems have gotten out of balance from this rapid change. Autoimmune diseases, those in which the body is attacking itself, have increased in prevalence at an alarming rate over the past decades, a rate that seems to track with the radical changes in our environment and lifestyle. In 1980, asthma had an incidence of 3.1 percent of the total population; today the incidence is 8.3 percent (a 268 percent increase), while other autoimmune diseases, such as multiple sclerosis, type I diabetes, and Crohn's disease, have experienced similar increases.[3]

Various theories have been proposed over the past decade to explain this drastic growth in autoimmune diseases. The hygiene hypothesis argues our immune systems are out of balance because we are no longer exposed to the plethora of viruses, bacteria, and parasites that thrived in pre-vaccine environments. In our modern world, certain parts of the immune system that normally would be hard at work fighting germs are underused, allowing other parts of the immune system, that normally would remain dormant, to inflame. The molecular mimicry theory, on the other hand, promotes the idea that new bacteria or viruses have a composition of proteins on their cell surface (antigens) that is similar to the composition of proteins on the surface of the body's own cells (self-antigens). When confronted with the new bacterium or virus, the immune system reacts appropriately by producing antibodies. But then, after the infection clears, the immune system continues its inflammatory reaction against self-antigens, mistakenly perceiving the body's own proteins as foreign.[4]

We can't predict what direction our immune system will decide to take next, with literally hundreds of different cell types interacting in a complex web of inflammation, suppression, and reaction. In our rapidly fluctuating environment, we are conducting a massive experiment on ourselves. Fortunately, knowledge is expanding and there is vast potential in future therapies emerging from the deeply intertwined disciplines of infection, asthma, autoimmune disease, and even cancer.

All of these fields of study are ultimately concerned with what our immune system is able to accomplish, and also what happens when the system goes awry. We are only now beginning to unlock some of the pieces, and lessons from asthma, particularly one discovered in the 1960s, show us both how far we've come while also sending us a warning about what could happen with our lungs in the future if we continue on the same path.

In 2005, during my second month as a pulmonary fellow at the University of Pennsylvania, I encountered a phrase that taught me more about asthma than any textbook. It doesn't mean anything out of context, but the four words *Get in your car* have stuck with me ever since I first heard them, and for me they reinforce the seriousness of an asthma attack.

Early on a Monday morning, I sat down at the computer next to my co-fellow, Mitchell. I had rotated off for the weekend the previous Friday night, and Mitchell was now eager to tell me about his weekend, and specifically about Mr. Nguyen, a young immigrant from Southeast Asia who had been admitted to the hospital that Friday night with a bad flare-up of asthma. Mr. Nguyen received the usual care—nebulizer treatments and steroids—and had initially gotten better. Then, at about ten o'clock that same Friday night, he got a whole lot worse.

The residents moved him to the intensive care unit, and his lungs continued to tighten up despite more nebulizer treatments. The resident called Mitchell at home about the case, and Mitchell then called the attending physician. The attending, Joseph, stopped him after about two sentences. "Get in your car. Right now. I'm heading downstairs and walking toward mine. Go and get in your car and meet me in the intensive care unit. Because this patient is going to need our help."

The patient was in *status asthmaticus*—asthma resistant to the usual medicines. *Status* comes from the Latin word *stare*, meaning *to stand*. The word *asthma* has Greek origins that go back thousands of years—indeed, it was mentioned by Homer in the *Iliad*—and originally meant noisy breathing. So this young man was *standing (still) with noisy breathing*. And if there's one thing the human body was not meant to do, it is stand still. The heart needs to beat, the kidneys need to filter blood, the gut needs to contract in peristalsis, the legs and arms need to move. And importantly, the lungs need to inflate and deflate to move air.

The attending turned out to be right; even with massive steroids and continuous nebulizers, the patient's lungs refused to loosen up. A decision was made to put him on a ventilator, and at around midnight a tube was placed into his lungs. Despite this, he continued to decline. They simply could not get his lungs to move enough air; they could not ventilate, and carbon dioxide was building up in his lungs and then his blood. Carbon dioxide in the blood naturally combines with water to dissociate into acid and bicarbonate. For this patient, the pH, a direct measure of acid in the blood, dipped from the normal 7.40 down to 7.13, and then 7.10, a dangerous level. As mentioned, the cells of our body, especially the cells that make up the muscle of our heart, cannot function properly in the presence of excessive acid.

Joseph, a few months out of fellowship, and Mitchell, a few months into fellowship, stood at the young man's bedside thinking about their next steps. They had swiftly reached the end of what is described in medical textbooks. So they did what many doctors do every day—they switched from *medicinae scientiam*, the science of medicine, to *ars medicina*, the art.

They tried some simple things first. Even though Mr. Nguyen was on a lot of sedative medication, his body was so revved up from respiratory failure that he was still fighting the ventilator, still trying to control his own breathing. So they gave him an agent to paralyze his muscles, letting the ventilator take over completely. Perhaps his body was fighting so hard it was impeding his ability to breathe, to exchange gas. After the paralytic, his body went limp, but a check of his bloodwork showed the dangerous level of acid in his blood hadn't changed. Mr. Nguyen's blood pressure then started dropping, because the pressure in his lungs was building up so high that it was squeezing his circulatory system.

With Mr. Nguyen's blood pressure crashing, Joseph and Mitchell tried one last thing—something Joseph called "the Australian maneuver," something not found in textbooks. They took Mr. Nguyen off the ventilator, then they both pushed down on his chest with all the force they could muster to squeeze all the air out. He couldn't get the air out of his lungs himself, so they did it for him.

At first this maneuver seemed to work, and Mr. Nguyen's air movement and blood pressure appeared to improve when they put him back on the ventilator. But then, over the course of five or ten minutes, the pressure in his chest built up again.

Joseph had one further idea but needed a surgeon to carry it out. He called the operator to page the cardiothoracic surgeon. "Stat!" he added. Two minutes later the surgeon called back, and Joseph spoke with him briefly. Soon after, the surgeon walked into the ICU with a package under his arm, and an associate wheeled a huge suitcase in front of him. Numbing the patient's neck, the surgeon plunged a large bore needle into the patient's jugular vein. Next, he pushed several plastic dilators into the vein to create a large hole, into which he inserted a catheter. They were setting him up to go on extracorporeal membrane oxygenation (ECMO), an artificial lung (described in Chapter 2) that oxygenates and ventilates the blood using a synthetic membrane housed in the suitcase they had brought with them. During the time ECMO is at work, the lungs can be partially deflated and lie idle, giving them a chance to heal. It doesn't always work, and problems with bleeding and infection can occur.

Fortunately for Mr. Nguyen, ECMO kept the oxygen flowing into his blood while removing the carbon dioxide and excess acid, and a week later his lungs were reinflated successfully. This time the pressure in his lungs didn't elevate, allowing him to first transition off ECMO, then off the ventilator, and eventually home.

For anyone who has experienced an asthma attack, or witnessed one firsthand, the sense of panic and urgency that it engenders is unforgettable. The immune system, when it believes it is under siege, can assemble very quickly. As if called to action, thousands of inflammatory cells are mobilized into the lungs in a matter of minutes, and a lot of fluid comes with them. An asthma attack can come on so suddenly, so ferociously, sometimes there is almost no time to react.

It matters which part of the lungs is attacked. Inflammatory diseases like pneumonia generally occur deep in the lung tissue, in the alveoli, where gas exchange takes place. This causes a problem with low oxygen levels in the blood, but providing extra oxygen can usually overcome the sensation of shortness of breath. Asthma, on the other hand, is known as an "airways disease," because the inflammation occurs at the level of the conducting airways, the bronchi and bronchioles. So when inflammation from asthma occurs, the movement of air is constricted, choked off. This creates the sensation of suffocation and is why sudden death can ensue. With this disease of the airways, oxygen is not the issue, since the gas exchange units of the lung, the alveoli, are unaffected. As in Mr. Nguyen's case, ventilation and carbon dioxide are the problem.

Sudden death from asthma is very much an issue today. In the United States in 2016, according to statistics from the CDC, there were 3,518 deaths from asthma, of which 209 involved children aged fourteen or younger.[5] While these numbers are not overwhelming, death from asthma should be almost completely preventable. The goal should be no deaths per year from asthma, especially in children, a group that very often has no other medical problems.

People with asthma have a lot of close calls. One of these stories is that of Javan Allison. Born prematurely, Javan struggled with breathing from day one. He was diagnosed with asthma at two years of age, and his mother, Monique, and father, Nick, quickly learned to give him breathing treatments and to recognize exacerbations. Later, Javan learned to use his own inhaler, and fastidiously gave himself his inhaled steroids twice a day. He did relatively well, but with his severe allergies to pollen, pets, and various foods, trouble was bound to happen, and his mother and father understood there would be days when an asthma attack might need more attention than they could provide at home.

Saturday, February 17, 2018, would be one of those days. Ten-year-old Javan woke up with some tightness in his chest and took his treatments as usual. Still constricted in the chest, he decided to take a shower, which in the past had helped loosen up his lungs. This time, however, things got so bad that Javan couldn't move enough air to talk. Within five minutes he had gone from a little chest tightness to being on the verge of asphyxiation. As a last resort, he stumbled out of the shower and wrote the words "HELP ME" on the steamed bathroom mirror with his finger.

His mother saw the writing and rushed him to the emergency room. Fortunately, a nurse who knew him from prior visits was there; nebulizer treatments were immediately hooked up, an IV was inserted, oxygen was placed on his nose, and medicines were started. After a few hours, Javan stabilized and was sent to the pediatric intensive care unit. On prior visits to the ER he had stayed a few hours. This time his breathing was so bad, Javan had to stay in the pediatric ICU for a week, his mother and father taking turns staying with him.

Today, Javan plays football, basketball, and baseball as well as studying math, his favorite subject. He does daily breathing exercises, continues to take his medicine every morning and evening, and always keeps an inhaler with him for emergencies. Javan and his mother decided to write a book about his experience. In it, Javan describes an asthma attack as being like "an astronaut flying out in space, but I have a crack in my astronaut helmet. As I get closer to the moon, the coughing gets worse. The crack in my helmet has gotten bigger. There is a pain in my chest as if a small meteor rock hit my helmet and broke it all the way open. Now, I'm in distress!"[6]

The book, the first in a trilogy, is called *The Adventures of Javan and the 3 A's,* the other two A's being anxiety and ADHD, which Javan also struggles with. No doubt these three conditions are linked. His and his parents' goal is to raise awareness, especially in the African American community, about these problems, and to remove any stigma about any of these conditions. With a changing environment and an immune system out of our control, feeling shame about asthma is the last thing people should have to worry about. And it's up to us to find out why kids like Javan are having to write "HELP ME" on bathroom mirrors.

Our basic understanding of asthma is that it's an allergic reaction focused in the lungs. Sometimes we are able to see allergic reactions, such as when poison ivy contacts our skin. Redness and irritation follow as the immune system is activated, along with extreme itchiness and swelling.

We put creams and lotions on the rash. After a few days, the swelling and itchiness go down, the inflammation melting away as magically as it appeared. Asthma is not so different, except it happens in the bronchi of the lungs, out of sight, somewhere more mysterious and inaccessible. And when the airways of the lungs get swollen, we can't breathe.

Descriptions of asthma go back thousands of years. Noisy breathing was noted in Chinese literature more than twenty-five hundred years ago. It was often treated with tea from the plant *Ephedra sinica*, which was probably somewhat effective. The active ingredient in ma huang tea, ephedra is a potent vasoconstrictor, helping the muscles that line our airways to tighten and let in more air during an asthma attack.

Accounts of noisy breathing and asthma also made their way into the medical literature of ancient Greece and Rome. The Roman philosopher Seneca experienced asthma firsthand, and in 62 BCE described it as "practicing how to die."[7] The Greek physician Aretaeus of Cappadocia also wrote about the struggle of an asthma attack, and in 100 CE produced this detailed account: "They breathe standing as if desiring to draw in all the air which they possibly can inhale; and in their want of air they also open the mouth as if thus to enjoy the more of it; pale of countenance except the cheeks which are ruddy; sweat about the forehead and clavicles; cough incessant and laborious; expectoration small, thin, cold, resembling the efflorescence of foam; neck swells with the inflation of the breath [pneuma]; the precordia retracted; pulse small, dense, compressed; legs slender; and if these symptoms increase, they sometimes produce suffocation after the form of epilepsy."[8]

Greek and Roman physicians didn't just document these attacks, they also had a list of cures, some of which had the potential to work. Foremost among them was the use of wine. One remedy was to mix ephedra with wine, another to add fox liver to wine. Other cultures have also struggled with asthma, each with unique cures. In India in the fifth century, smoking stramonium, a plant of the nightshade family that has muscle-relaxant properties, was popular. In the Americas, cocaine, tobacco, and balsam have been used to help open up the airways.

One of the more famous sufferers of asthma was novelist Marcel Proust, who lived in Paris at the turn of the twentieth century. Beginning at age nine, he was plagued by attacks of breathlessness, many of which he wrote about in his later life. In 1900 he wrote to his mother: "An attack of asthma of unbelievable violence and tenacity—such is the depressing balance sheet of my night, which it obliged me to spend on my feet in spite of the early hour at which I got up yesterday."[9]

Proust went to great lengths to control his asthma, including frequently smoking stramonium cigarettes imported from India. In addition, Proust also tried morphine, opium, and a variety of inhalants. He flipped his days and nights to escape the worst of the pollen from trees and flowers, never opened his windows, and lined the walls of his room with cork paneling to keep out bad humors. He did most of his writing at night, securely isolated in the stillness and quiet, hoping his lungs stayed calm.

In Proust's day, the cause of asthma was unknown, and different theories abounded. It was considered to be a disease of upper-class city dwellers, perhaps the reason it was looked upon as a badge of honor and civility (the opposite perception prevails today). Asthma in women was seen as having a hysterical component rooted in a psychiatric condition that could potentially be counseled away by Freudian methods.

Some scientists attempted to isolate different causes of asthma. Dr. Morrill Wyman of Harvard University noted that his allergies and asthma melted away when he left the Boston area and visited the White Mountains of New Hampshire. As an experiment, he took a sample of ragweed from Boston with him one weekend to New Hampshire, and upon inhaling its scent he was "seized with sneezing and itching of nose, eyes, and throat with a limpid discharge. My nostrils were stuffed and my uvula swollen, without cough, but with the other usual symptoms of Autumnal Catarrh." Dr. Wyman published his observations in an August 1875 article in the forerunner to the *New England Journal of Medicine*, and he would go on to distribute the world's first pollen maps to help travelers with asthma plan irritant-free vacations.[10]

While it became clear that different allergic stimulants in the atmosphere could trigger asthma, what exactly was happening in the bodies of some people to convert this exposure into a near-death experience was not known.

The beginning of the twentieth century has been referred to as the first time when seeing a doctor might actually be beneficial. This was true with respect to asthma, and the main reason was the discovery of the immune system, the small cells in our blood that react to external stimuli for better or for worse. Allergy, asthma, hypersensitivity, and infectious disease are all different expressions of how the immune system reacts to external stimuli. Significantly, cancer has become part of the story of the immune system, as we recently have come to appreciate that our immune cells have the ability to recognize cancer cells as foreign, and then attack them, much as they do a virus or bacteria.

We first began to learn about the immune system in the second half of the nineteenth century. White blood cells are the main ammunition utilized by the immune system, and there are several different types of them: B lymphocytes, T lymphocytes, and mast cells, along with neutrophils, basophils, and eosinophils. All of these cells are programmed to react to different invaders, aiming to protect us by killing bacteria, viruses, and parasites. By the time the nineteenth century was drawing to a close, scientists had observed that these white blood cells were present in abundance wherever there was a bacterial infection, an asthmatic reaction, or some other immune stimulus in the body. But many questions remained, including which cells got stimulated under different circumstances, and whether they retained any protective memory of past insults.

One of the first clues to how the immune system functions came from the fight against diphtheria. The bacterium *Corynebacterium diphtheriae* was a scourge in the late nineteenth and early twentieth century, frequently infecting children and resulting in a 20 percent death rate. At the time, it was the third leading cause of death in children in Britain. Known commonly as the "strangling angel," it got the name *diphtheria* from the Greek word for *leather*, because of the thick gray inflammation it caused in the back of the throat, which in turn caused respiratory failure by choking off the airway.

A breakthrough came in 1891, when the German scientist Emil von Behring began experimenting with the serum component of animals that had contracted the disease. Serum is the portion of the blood without the red and white cells, a yellow-colored liquid containing mostly proteins. Von Behring noted that the serum from animals that had survived diphtheria could be transferred to another animal struggling with the disease, resulting in a cure. Humans were next. As the story goes, on Christmas Eve in 1891, von Behring used animal serum to successfully treat his first patient, a little girl in a Berlin hospital who otherwise surely would have died. Later, other scientists helped von Behring concentrate the animal serum, and further use in children produced miraculous turnarounds, most within twenty-four hours. Suddenly, the dying children who were admitted to hospitals with the gray inflammation in the back of their throats were surviving.[11]

Horse-derived-serum therapy swept through Europe and on to America. Horses were purchased by the New York City public health department and kept in a stable on the Upper West Side to produce a steady stream of serum for the five boroughs. The results were dramatic: the number of diphtheria-related deaths in New York City dropped from 2,870 in 1894 to 1,400 in 1901. The stable was eventually moved to East Fifty-Seventh Street, and then to upstate New York, where antitoxin was produced for the next several decades until a vaccine was developed. (Serum therapy most recently made a comeback with the COVID-19 crisis, the treatment essentially unchanged since first used by von Behring in 1891.[12])

Serum therapy was important because it proved that white blood cells were not fighting bacteria alone; some antitoxin factor in the blood, separate from the white blood cells of inflammation (our lymphocytes, neutrophils, eosinophils, etc.), could control disease. But exactly what it was in the serum that was doing this work was unclear, and soon after the use of serum became widespread, a disturbing thing about our immune system emerged. The scientist who first noticed this disturbing thing was Clemens von Pirquet, a Viennese pediatrician born in 1874. Von Pirquet's ideas on immunology, like most revolutionary ideas that end up being correct, were largely criticized, dismissed, and ridiculed by his peers at the time.

Von Pirquet was first a clinician and second a researcher. His primary job was at the Children's Clinic in Vienna, where he observed something interesting in some of the patients who were treated for diphtheria with the serum from a horse. Most of the children who received antitoxin did well, but a stubborn minority did not. In fact, some got even sicker and died more quickly than expected.

Von Pirquet's observations led him to introduce the radical idea that our immune system itself could be a problem. These children's immune systems appeared to recognize the horse serum as foreign and were reacting in a manner extreme enough to cause death. Before von Pirquet, doctors believed that disease was the fault of whatever invaded a patient's system. Von Pirquet's observations seemed to indicate that the immune system itself could go out of control.

This new way of looking at the problem prompted von Pirquet to create a new word: *allergy*, from the Greek *allos* (other) and *ergos* (activity). Von Pirquet's concept seemed ridiculous to many at the time, and even to himself. In 1906, he remarked that "the concept that serum, which should protect against disease, is also responsible for disease, sounds at first absurd."[13]

Other observations not only reinforced von Pirquet's idea that a person's own blood could be destructive, but also that this phenomenon was linked to asthma. One of the most dramatic examples came in 1919, when an immigrant from Greece, identified in the medical literature only as H.T., arrived in Central Park in New York for a relaxing stroll and a horse carriage ride. As soon as he entered the carriage, he was struck by a severe asthma attack. He was stunned. He had been around horses many times without an issue, and he had no history of asthma. To his shock, when he went back to Central Park the next day, the same thing happened.

H.T. went to see Dr. Maximilian Ramirez, who carefully asked his patient about anything unusual that had occurred in his life recently. After hearing the answers to his questions, Dr. Ramirez not only had a diagnosis, but he had the basis of an extraordinary research paper and a place in the annals of the history of immunology.[14]

The patient H.T. told Dr. Ramirez little of importance, except that he had received a blood transfusion two weeks earlier. This was enough for Dr. Ramirez to formulate his ideas. He tested H.T. for allergies, and the patient came up positive to horses. Pushing forward with his theory, Dr. Ramirez tracked down the blood donor, who not only admitted to being a lifelong asthmatic and allergic, but had a horse allergy test that was even more positive than H.T.'s. Dr. Ramirez correctly surmised that some sensitizing agent had been passed from the donor to H.T. during the blood transfusion. Here was the first definitive case showing that allergies and asthma could be passed from one patient to the next through the blood.

Still, the question remained whether the white blood cells or some other factor in the serum was the initial trigger of allergy and asthma. The answer came three years later, in 1922, with the experiments of Drs. Carl Prausnitz and Heinz Küstner in Poland. Küstner had many allergies throughout his life, including a severe fish allergy, while Prausnitz was allergy free. As a test, they took some of Küstner's serum and injected it into Prausnitz's arm. The next day they injected a fish extract into Prausnitz's arm at the same site where Küstner's serum had been inserted, and his arm shortly developed a swollen, itchy, red rash. For the first time in his life, Prausnitz had developed an allergic reaction.

Now the cause of allergies and asthma had been narrowed down to the serum. Some sensitized factor or protein was reacting to foreign stimuli and causing the body to inflame. Prausnitz and Küstner coined the term *reagin* to describe this factor in the serum, but they really had no idea what it was or what comprised its biochemical makeup. Almost fifty more years passed before this was figured out, and like many scientific discoveries, chance played a huge role. Indeed, two separate teams of scientists happened to make the discovery at exactly the same time using two completely different methods.

The field of immunology advanced steadily over the five decades after Prausnitz's reaction, but the element in the serum responsible for allergy remained unknown. Researchers could name the different elements that triggered inflammation, whether it be grass or pollen or ragweed or an animal. But they had no idea what these allergic proteins were binding to in the serum that was triggering the stuffy nose of allergies or the deadly constriction of asthma.

An important advance came in the 1950s, when scientists gained a greater understanding of the proteins in our serum, some of which were antibodies. The term antibody had been used in the late nineteenth century, but the structure of the proteins had not been elucidated. Shaped like Y's, antibodies attach to foreign particles with the open end, triggering other immune cells to migrate and attack, attaching onto the straight portion of the Y. This complex in effect neutralizes whatever the foreign particle is, usually a bacterium or a virus, but inflammation and fluid frequently come along with the process, for better or worse. Also called immunoglobulins, the antibodies were given different letters based on what their role was thought to be. For example, immunoglobulin G (IgG) is an antibody believed to play a big part in fighting off infections. It likely played a role helping those with diphtheria fight off the bacteria. Scientists tried tying these different antibodies to allergy and asthma, but without success, until the 1960s.

The search for the structure of Prausnitz and Küstner's reagin began in two separate laboratories—one in Denver, Colorado, and the other in Sweden. The two labs took completely different approaches to the problem, but finally, in a moment of collaboration between scientists and institutions, they came together to fit the final pieces of the puzzle together.

By the time he moved from Japan to Colorado in 1962, Dr. Kimishige Ishizaka had been working on discerning the structure of reagin for some fifteen years. He had made some progress but had also suffered some demoralizing setbacks. He had tested one of the known antibodies, immunoglobulin A, excitedly thinking it might be the source of reagin, the lynchpin of allergic disease, but his experiments ended up failing.

Dr. Ishizaka began a series of trials with rabbits, injecting them first with serum from a patient with severe allergic disease. The rabbits' bodies recognized the serum as foreign and created antibodies to the proteins in it. Presumably, one of the antibodies the rabbits would make would be in reaction to reagin, sure to be contained in the allergic patient's serum. To isolate the anti-reagin antibodies and reagin itself (which would be bound to the antibodies), Dr. Ishizaka took the rabbits' serum and washed it of all known proteins.

To test his belief that, after washing this sample, he now had antibodies to reagin, Dr. Ishizaka took more serum from an allergic patient and added his purified rabbit serum. When he added an allergy-generating molecule into the mixture, the allergic patient's sample displayed no sensitivity—the newly created rabbit antiserum had obstructed any allergic inflammatory reaction. Dr. Ishizaka now clearly had isolated antibodies to reagin, and reagin itself. He renamed reagin yE-globulin, and he suspected it was a protein closely related to antibodies. This was a huge breakthrough, but he still had not determined the exact structure of the protein he had isolated, until he received a letter from Sweden.[15]

At the same time as Dr. Ishizaka was undertaking his trials, a group in Uppsala, Sweden, led by Gunnar Johansson and Hans Bennich, had become interested in a protein from a patient with the blood cancer multiple myeloma. Myeloma is a blood cancer disease of B cells, the cells that produce immunoglobulin. One patient with obvious signs of myeloma had a protein in his blood that wasn't one of the known immunoglobulins. With advanced molecular techniques, the Swedish group began analyzing this unique protein. (They had the benefit of having the protein in far higher concentrations than Dr. Ishizaka did, since their patient with cancer was producing the immunoglobulin unchecked.)[16]

Their protein analysis confirmed it was an immunoglobulin, a new one with a structure far smaller than other immunoglobulins. They named their new protein IgND, incorporating the initials of the cancer patient. They suspected IgND was also reagin, especially after reading Dr. Ishizaka's paper on his efforts to isolate the protein and its antibodies. Johansson and Bennich mailed a letter to Dr. Ishizaka to confirm their thoughts.

The two groups exchanged their samples and established that the proteins were the same. The following year, in 1968, they collectively renamed the protein immunoglobulin E (IgE). The official lynchpin of asthma and allergy had been isolated after a fifty-year search. Part of why it took so long, and why Dr. Ishizaka's initial isolation is so remarkable, is that reagin, or IgE, is present in the blood in concentrations of about ten thousand times less than other immunoglobulins, even in those with severe allergies.

We now understand that when some foreign protein such as ragweed is inhaled into the lungs, the antibody IgE binds to it, triggering an inflammatory cascade that includes an influx of other white cells, especially the eosinophil and mast cell. The allergic protein-IgE complex then binds to the eosinophil or the mast cell, activating them to undergo a process of degranulation, whereby other inflammatory proteins contained within the eosinophil and mast cell are dumped into the blood, attracting many more white blood cells in a cascade of inflammation. Fluid then follows the inflammatory cells, in turn causing airway swelling, which leads to constricted bronchial tubes and difficulty with breathing and ventilation. Only when the threat appears to have passed do the inflammatory cells leave the lungs, the fluid gets reabsorbed, and the bronchial tubes return to their normal caliber. Steroids, inhaled and intravenous as needed, are anti-inflammatory, helping to suppress and drive out the eosinophil, the mast cell, and other inflammatory white blood cells from the lung. Bronchodilators, such as albuterol, aid as muscle relaxers, easing the constricted bronchial walls open.

This, at the molecular level, is what is happening during an asthma attack, and while it usually develops over several days, it can happen quite suddenly. For some reason, allergic patients have highly sensitized IgE and misrecognize harmless proteins from grass and dog fur and cat hair as deadly, while IgE in those without allergies stays dormant and in small amounts. Sensitized IgE, likely in large amounts, is what was transferred from the asthma patient to H.T., and from Küstner to Prausnitz, triggering both of their allergic reactions.

Later research revealed that the proper role of IgE and its accompanying white blood cell, the eosinophil, is to fight infections, specifically parasites. But why this reaction crossed over to things like pollen and dog hair is still completely unknown. We do know this crossover reaction is happening nowadays with alarming frequency, perhaps because IgE has so much less work to do on parasites.

Soon after its discovery, a test was developed to measure IgE in the blood, allowing doctors to get a precise measurement of the level of allergic inflammation in their patients' bodies. Later, it became possible to measure not only total IgE but also how much of it was sensitized to different allergens, such as pollen, ragweed, trees, grass, or cats. For the first time, patients could find out with a simple blood test not only what they were allergic to, but how severe their allergies were. This allowed them to take preventive measures, some as simple as vacuuming more to get rid of dust mites, or maybe avoiding the cat or dog triggering their chest tightness.[17]

Knowledge of IgE also led to improved treatments. In 2003, the medicine Xolair was approved as a remedy for severe asthma patients. An antibody itself, the drug binds IgE in the blood, giving asthma patients with severe disease an option beyond steroid-based inhalers. Steroid inhalers target the eosinophil and other white blood cells, but Xolair is even more specific and attacks at the initial focus of inflammation. Over the years, it has been shown to improve lung function, reduce hospitalizations and ER visits, and improve overall quality of life. It can also help patients reduce their reliance on steroids, which can have unpleasant side effects even in the inhaled form.[18] Other antibodies similar to Xolair have since been developed and approved as our understanding of the immune system expands.

*　*　*

Even as our understanding improves, the human immune system continues to evolve, largely out of our control. We have removed ourselves from a multitude of dangers, but the immune system changes and develops deep within us. Lately, it has turned its attention to attacking its own host, unclear about its role, hitting us with a wave of asthma and allergic disease like we have never seen before. We know that IgE is a large part of what drives asthma, but why so many more people are getting sensitized IgE that reacts with typical allergic particles is unknown.

Some of this should come as no surprise, as changes to our environment over the past century, even the past decade, have been extreme. With rising carbon dioxide levels, there is in fact more vegetation on Earth today than a decade ago, as plants utilize the increasingly abundant CO_2, and humans plant more to offset effects of CO_2 production.[19] Known as "global greening," the increased vegetation has brought ever more pollen, and with it an expanding allergy season and more severe symptoms for those affected.

Other changes being navigated by our immune system include worsening air pollution and a shifting landscape of infectious diseases. There is no apparent end in sight to the increasing asthma rates and severity of disease, along with the skyrocketing prevalence of a multitude of autoimmune diseases. Despite the complexity of the immune system, the question that needs answering is likely quite simple, and goes back to our lungs: What are we breathing—or not breathing—today compared to a generation ago that is driving this massive increase in inflammation.

Chapter 6

The Lungs and
the Common Good

Eduardo Rosas Cruz was arrested late in the evening near Bakersfield, California, on July 28, 2014. He didn't assault anybody, didn't steal anything, and wasn't driving under the influence. He wasn't trespassing or jaywalking. In fact, he got arrested for something even the best-intentioned and hardest-working of us often forget to do: he didn't take his medications.[1]

This arrest would appear to violate the medicolegal principle that a person of sound mind has the right to refuse any treatment or medication. But the case of Rosas Cruz was not typical. His lungs were infected with tuberculosis, and with the communal nature of our atmosphere, the state of California was within its legal rights to imprison him. The prosecutor, Stephen Taylor of San Joaquin county, put it succinctly: "Criminal prosecutions (can be) an extension of the practice of medicine."[2]

The arrest of Rosas Cruz represents a rare intersection of crime and medicine. It is an extreme example, but it serves to demonstrate how the lungs exist at the juncture between our communal air, deadly infections, and our rights as citizens. Today, the intersection of these three issues is becoming more fraught. With growing urbanization, extreme mobility, and a plethora of new bacteria and viruses emerging, lung health serves as a bellwether for what is happening in society worldwide. The 2019-2020 emergence of the novel severe acute respiratory syndrome coronavirus 2 (SARS-CoV-2, commonly known as COVID-19)—with the extraordinary subsequent shutdown of daily life, closed borders, and overwhelmed medical systems—is the perfect example of how a lung infection can take over society in a very short period of time.

The connectivity of our air can be hard to conceptualize. We can easily imagine the oceans as being connected: when a shipping container dumped thousands of Legos off the coast of Cornwall, England, in 1997, we were not surprised that some pieces were later recovered in Ireland, Galveston, Texas, and Melbourne, Australia. We can see the ocean currents. However, the air of our atmosphere, while invisible and intangible, is no less connected than our oceans. This is true at the local level, where infections can spread easily from lung to lung, but also at much greater distances. A recent study has shown the trees in Yosemite National Park are reliant on nutrients from the dust in China's Gobi Desert that have caught a ride on the east-west jet stream, traveling some six thousand miles.[3] Similarly, the Amazon relies on the dust of the Sahara Desert in Africa, which travels in the opposite direction but just as far.[4]

For humanity, the world that is our atmosphere has gotten smaller, more communal. The unseen air that is becoming increasingly more collective is something that needs to be respected, and the threats to it must be addressed at local, national, and international levels. No other organ teaches us this lesson better than the lungs, and no other diseases better than tuberculosis and COVID-19.

In November 2005, during my first year of pulmonary fellowship, I fielded a call about a sick patient who was on the general medical floor. He was twenty-two years old, a college kid who had come in with a fever and what looked like pneumonia, but he wasn't responding to typical antibiotics. At night his fever spiked upward of 103 degrees Fahrenheit, and sweat poured out of him, saturating his bed linens. He had lost a great deal of weight, and his chest X-rays showed more inflammation every day.

We had no idea about a diagnosis, so we took him to the surgical suite, sedated him, and inserted a camera into his lungs, taking some small biopsies, tiny pieces of lung tissue, in the hope of finding a diagnosis. In retrospect, it is clear that the answer was right in his history, as almost all diagnoses in medicine are.

We learned that our patient had done something over the summer that thousands of young adults throughout the country had done—he had gone to an outdoor rock concert, this one with almost one hundred thousand people, all sharing the same cloud of air. Music festivals are among the most likely venues for infectious diseases to do their work: a lot of people from all over the world coming together into a tight space with variable sanitation facilities. Some cities, especially a hundred years or more ago, could also be described in this manner.

We did our bronchoscopy, got our tissue, and soon thereafter had our diagnosis: the young man had tuberculosis, and a lot of it. After his diagnosis, the patient was kept isolated in his room so he couldn't inadvertently spread his bacteria, and per protocol and law we reported the disease to the Philadelphia public health department. Everything seemed to go smoothly; our young patient started his medications, his fever went away, and his cough got better. He was sent home, and every week he picked up his pills at a city-run clinic. He seemed reliable, and stated to the outpatient clinic that he was taking all his medications without an issue.

At home, however, the patient's fever came back, as did his cough and shortness of breath. He came to the emergency room, where his breathing was found to be so bad, he ended up with a tube down his throat and on a ventilator, and was admitted to the ICU. Once we stabilized him, everybody involved tried to figure out what had happened to someone who had taken his medications and seemed on the road to recovery.

We got a CT scan (a more detailed X-ray), which showed that the inflammation in his chest had increased—after a month of treatment. Even worse, when we did another bronchoscopy, we saw that he still had TB organisms actively growing in his lungs.

We followed up with the health department about his bacterium to make sure it was susceptible to the antibiotics we were giving him. The health department in every state not only records every case of TB, but also reviews drug sensitivities (how well a strain of bacteria responds to medications). We were assured his strain of tuberculosis was sensitive, and that our patient had been taking his medications. We then checked him for basic immune deficiencies, and again nothing turned up.

With the obvious accounted for, we turned to the obscure, the area of medicine where judgment and experience come into play. Fortunately, we received sound advice and guidance from the infectious disease physicians. Stick to the basics, they stressed, only do them better. We continued our patient on the first-line anti-TB drugs he had been on, this time at a slightly higher dose, and added one dose intravenously since he had inflammation in his abdomen and likely wasn't absorbing a lot of the medicine he had been taking. We supported his immune system with appropriate calories through a feeding tube and kept the pressure in his lungs low on the ventilator. In this case, there could have been an urge to change the plan radically, to alter the anti-TB drugs or give steroids or other immune modulators. The art of medicine is knowing when to give up and start anew versus when to stay with the basic plan, executed better. In this case, we stuck with the basic plan.

Gradually, the inflammation in the chest and abdomen calmed down. We were able to give the patient physical therapy, and he got off the ventilator and went to the general medical floor. We stuck to the basics of antibiotics and nutrition, and he did well. When in doubt in medicine, *stick to the basics* are indeed words to live by.

Although we didn't recognize this until quite recently, infectious organisms have been spreading through the air and gaining entry into our lungs for centuries. Lung diseases such as influenza, anthrax, measles, and tuberculosis have all had a vast and wide-ranging effect on human life and culture. Newer illnesses, such as the severe acute respiratory syndrome (SARS) virus in Asia, and the Middle East respiratory syndrome (MERS) virus in Saudi Arabia, have also been shown to spread through the air, as has the COVID-19 virus from Wuhan, China, that first emerged in December 2019. The atmosphere is a communal space, and the lungs are an extension of it. Bacteria and other organisms have exploited this shared space for centuries to jump between people and multiply unseen.

This is not news to us, and most of us recognize that a potentially lethal disease could emerge from another person's lungs and make its way into ours. If somebody coughs on the subway, that hacking will send everybody to the other side of the car. We are told to cover our coughs, to sneeze into our elbow, and then to wash our hands afterward. The lungs, mostly unrecognized for the work they do, are best known in modern culture as the petri dish and transmission vehicle for deadly diseases.

The science behind what happens in the air when we cough, sneeze, or talk has recently been examined, but there is still much we don't know about how our secretions travel or how bacteria and viruses take advantage of different environmental conditions to move from host to host. Much of the original research into this subject was done in the late nineteenth and early twentieth century, when air was recognized as the vehicle for contagions; without any effective medicines or vaccines, this was one area that public health experts could focus on to control disease. With the COVID-19 crisis, much of this research is being revisited, and interest in the mechanics of contagion spread has increased.

One of the first people to demonstrate the transmission of organisms in the droplets of our secretions was German infectious disease doctor Carl Flügge. In 1897, Flügge placed samples of the harmless bacterium *Bacillus prodigiosus* into the mouths of volunteers and then documented the presence of the bacterium in droplets around those subjects after they talked or coughed.[5] Today, droplets are defined as molecules that are greater than ten micrometers in size and contain the intact fluid from our secretions as well as potentially infectious organisms. Due to their size and weight, they generally travel no more than six feet after being expelled from our bodies and then settle on nearby surfaces. Droplets can remain infectious for hours to days, but they spread primarily by somebody touching the expelled secretion and then touching their own mouth or nose; the risk of directly inhaling droplets is usually limited to a six-foot zone. Given their relatively large size, the droplets themselves are filtered out prior to reaching the lungs, but the organisms contained in a droplet can spread to the lungs after replicating in the nose or throat.

In the 1930s, Dr. William Wells of Harvard University took the research of contagion spread a step further. At his laboratory, he built a chamber in which he atomized different liquids, and by projecting a strong beam of light he demonstrated their quick dispersion. He then added bacteria to the droplets, and while some, such as *Staphylococcus aureus*, disappeared quickly from the air, others, such as *Bacillus subtilis*, persisted after three days, which should not have been possible according to the current droplet science. Dr. Wells also conducted unique experiments at the Harvard School of Public Health, introducing sneezing powder into a lecture hall and then collecting the normal bacteria of his graduate students that was dispersed throughout the room. He also inoculated the air conditioner in the basement with *Balantidium coli*, and later recovered the organism from every corridor of the three-story building.[6] With his findings, he promoted the term *droplet nuclei*, smaller molecules, less than ten micrometers in size (modern definitions may use less than five micrometers), in which the fluid has largely evaporated from the droplet, leaving an infectious particle that does not fall to the ground. This particle could clearly stay suspended in air for long periods of time, often traveling far afield. Given their smaller size and their potential to stay suspended in air for hours, droplet nuclei can easily be inhaled and make their way into our lungs, avoiding the filtering system of our nose and throat, and begin replication, initiating pneumonia directly.

The distance traveled by droplets and droplet nuclei depends on several factors: the location of the person emitting them (indoors or outdoors); the environmental conditions of temperature, humidity, and ventilation; as well as how the secretions are initially expelled from the person. Sneezing is the most powerful form of exhalation, and a single sneeze can produce as many as forty thousand droplets traveling at speeds of one hundred meters per second. A cough may expel three thousand droplets, while simple talking for a minute produces about six hundred droplets.[7] Coughing and sneezing may also produce gas clouds, and a study in 2014 from MIT showed these clouds of gas can travel much farther afield than was initially suspected, easily making it into the ventilation units of a room.[8] Different procedures in the hospital create their own unique infectious risks, such as when cardiopulmonary resuscitation is undertaken and somebody is pressing vigorously on the chest, or when a tube is inserted into the lungs to place a patient on a ventilator.

Respiratory infections today are generally broken up into those that spread through droplet transmission via inhalation at close proximity or contact with droplets on surfaces, and those that are able to survive in droplet nuclei and spread through strictly airborne means. Influenza and COVID-19 are believed to use droplets to spread, while organisms such as tuberculosis and the measles virus spread more through droplet nuclei.[9] This influences the type of protective barrier that is necessary, with a six-foot separation and a mask that can stop relatively large particles recommended for COVID-19, while those with tuberculosis need to be kept in a negative pressure room to continuously suck out the air, and those taking care of them need a tight-fitting mask that has the ability to trap small particles.

However, given the many factors influencing the spread of a virus, such as particle composition, mode of being expelled from the body, and specific environmental conditions, the act of staying six feet away from a person infected with influenza or COVID-19 may not always be protective. A sneeze can travel up to twenty-seven feet, and a gas cloud from a cough can extend the life of a droplet from a fraction of a second to minutes. A 2020 report from China bears out this concern. According to this report, particles of COVID-19 were found in the ventilation systems of the hospital rooms of those infected.[10] Based on the principle that droplets fall to the ground within six feet, this should not have been possible. We also have no clear understanding of how long protective equipment such as masks can last and under what circumstances, and if cleaning them affects their ability to protect. The eyes are another potential route of transmission. How much spread occurs through secretions coming in contact here is unknown, but accounts of an initial eye infection as the source of illness have been reported in those with COVID-19.[11]

The movement of the air in our atmosphere can be very beneficial, by transporting plant seeds and nutrients, and by diluting out toxins and smoke. Not surprisingly, bacteria and viruses have learned to exploit this system, specifically to travel from one host to another. The air is indeed collective, with us often unaware of unseen and unseeable threats.

In the first half of the twentieth century, there weren't many specialists in medicine; but starting around 1950, with the development of new technologies, kidney doctors, heart doctors, and brain doctors all started carving out niches for themselves. For lung doctors, while little new technology existed, many cases of tuberculosis did. In fact, at the beginning of pulmonary specialization all lung doctors were TB doctors.

TB has been with humanity for so long, and its history with us is so extensive and varied, an argument could be made that no other infection or disease has affected humankind to a greater extent. It has touched many aspects of our culture, from our novels to our paintings to our people. If the history of civilization is surveyed in its entirety, no other infectious disease has killed more people than TB, over one billion in the last two hundred years alone.[12] And it continues to kill more than one million people worldwide each year.[13]

Much as civilization itself, tuberculosis first appeared in East Africa some twenty thousand years ago.[14] It has stuck with us ever since, and today it exists in the latent stage in almost two billion people, a quarter of the world's population. Being in the latent stage means TB infected these people at some point, was controlled but not completely eradicated by the inflammatory system of their lungs, and has the potential to reactivate if their immune system weakens.

Looking back, we have a detailed historical record of TB's existence. In analysis of ancient Egyptian mummies, paleopathology (medical analysis of ancient dead bodies) shows that some of the priests and priestesses were afflicted with TB. In 1891, in the ancient city of Thebes, forty-four well-preserved mummies were discovered dating back to about 1000 BCE. One, named Nesperehan, was an adult male with partial destruction of the lower thoracic and upper lumbar vertebrae, which created an acute angular deformity. Paleopathologists pin this type and location of spinal destruction clearly on tuberculosis.[15]

Modern techniques of DNA analysis have confirmed the existence of TB in precolonial South America as well. Surprisingly, the DNA imprint of the TB found in South America is not close to the typical European or African TB, but rather resembles the TB found in seals.[16] The researchers' conclusion was that seals brought TB to the Americas, picking it up from Africa and then swimming across the Atlantic and spreading it to those who hunted them along the coastline of South America.

Over the centuries, tuberculosis continued its march through civilizations, with names like *phthisis* in ancient Greece, *tabes* in ancient Rome, and *schachepheth* in ancient Israel. If the prevalence of tuberculosis ebbed slightly during the Middle Ages, that decrease was followed by a huge increase in Europe and North America during the eighteenth and nineteenth centuries. With more of the population concentrated in urban environments, and no detailed understanding of TB's cause or mode of transmission, up to 90 percent of residents in certain cities got infected. For a time in nineteenth-century Europe and America, TB accounted for one out of every four deaths.

During this two-hundred-year period, TB became known as the "white plague," due to its slow-acting way of sucking the life, weight, and health out of people, turning a patient from a person into a ghost. Painters and other artists began romanticizing the disease. The novelist Amantine-Lucile-Aurore Dupin (better known as George Sand) referred to her lover, the composer Frédéric Chopin, as "a poor melancholy angel," and described how he "coughed with infinite grace." The British poet Lord Byron, in 1828, wrote, "I would like to die from consumption. The ladies would all say, 'Look at that poor Byron, how interesting he looks in dying!'"[17]

From Mimi, the heroine in Puccini's *La Bohème*, to characters in Eugene O'Neill's plays and Fyodor Dostoyevsky's novels, tuberculosis played a big role in opera, literature, and other arts. One of the most fascinating case studies of TB in art, and its effect on the artist and his career, involved Edvard Munch's painting *The Sick Child*, depicting his sister Johanne Sophie, who died at age fifteen from TB. Best known for his later Expressionist classic *The Scream*, Munch had an artistic breakthrough while working on *The Sick Child*: "I started as an Impressionist, but . . . Impressionism gave me insufficient Expression—I had to find an expression for what stirred my mind . . . The first break with Impressionism was *The Sick Child*—I was looking for expression."[18]

The painting has two people in it, but only one face, that of Munch's sister, a young girl with red hair and a caved-in chest, lying in bed and looking with yearning toward an older woman sitting in a chair next to her. The older woman is trying to comfort her, and their hands are symbolically intertwined, without much definition, symbolizing the meshing of their souls.

The older woman represents Munch's aunt, Karen Bjolstad, but her face is down, completely unseen. The tragedy of the moment appears to weigh too heavily on her heart, the knowledge of what the child is going through and what she is facing with an incurable disease that is too much for her to handle. Perhaps this woman is also partly Munch himself, faced with the guilt of surviving tuberculosis as a child while his sister did not. It's also obvious, to those who know Munch's story, why his aunt and not his mother is holding his sister's hand: his mother had died when Munch was six years old, also of tuberculosis.

The Sick Child is one of the most famous of Munch's paintings, the canvas thick with broad paint strokes and emotion—greens and blues, representing illness and sadness, but also spots of red representing the deadliness of the disease, as well as its propensity to cause bloody sputum. It's a painting that Munch not only worked on for a year, but one that he came back to many times to repaint and reimagine over the course of forty years. When he moved to Paris in 1896, he repainted *The Sick Child* several times in different colors, and later painted four more versions of it, two in 1907, one in 1925, and the last one in 1927, when Munch was sixty-two. The scars of the illness on his family, and on him, would not heal during his lifetime, the canvas becoming his medium for therapy.

Scientists have debated the cause of tuberculosis for millennia. Hippocrates believed the disease to be hereditary, because so many within a family seemed to contract it. Galen, a few hundred years later, believed it to be contagious, as well as incurable. Later, in 1546, the Italian Renaissance physician Girolamo Fracastoro wrote insightfully that TB is caused by "*seminaria contagiosum*," or infectious seeds, and that the bedsheets and clothing of TB patients could be highly infectious. Galen's miasma theory of disease, in which diseases were thought to originate in the air from normal rot of organic matter, also continued to be a popular explanation during this time.[19]

The case for TB as an infectious agent did not gain firm scientific support until the mid-nineteenth century. The first breakthrough study was done by French military surgeon Jean-Antoine Villemin, who noticed that recruits kept in the barracks were much more likely to contract the disease than those in the field. To test his hypothesis, he cut out an inflammatory lesion filled with pus from one of his patients who had died from TB and showed successfully that a rabbit exposed to this substance would develop the disease. He published his findings in 1865, in a paper entitled "*Cause et nature de la tuberculose: son inoculation de l'homme au lapin*," or "Cause and nature of tuberculosis: its inoculation from a human to a rabbit."[20]

Villemin was mostly disregarded, partly because he had only described the lung pathology of the disease and shown that it could be spread from person to person. He suspected the cause of TB was a bacterium, but he did not isolate the organism. What the world needed was to see the bacteria, evidence of which Robert Koch finally produced seventeen years later. With his findings and methods, Koch also laid the groundwork for the modern study of bacteria and for the widespread adoption of the germ theory of disease. His discovery received so much press, at a time when TB had such a strong hold on society's consciousness, that it had wide-reaching implications for how we view disease, and how we function as a society.

Born in Hanover, Germany, in 1843, Dr. Koch excelled at his studies, graduating in 1866 from the Göttingen medical school with the highest distinction. He married, had a daughter, and then worked as an army physician during the Franco-Prussian War in 1872. After the war, he settled in Wollstein, a part of modern Poland, and opened a medical practice. As a gift for his thirtieth birthday, his wife gave him a microscope, which he immediately used to study the bacterium anthrax. Despite the heavy demands of a practice, he built a laboratory in his house and set to work proving that anthrax was the cause of the disease afflicting local farm animals. He subjected mice to the infected blood of sheep, then documented their subsequent illness and death. Afterward, at autopsy with the use of his microscope, he recorded evidence of the small, rod-like bacteria in their blood, spleen, and lymph nodes. Although in retrospect his experiments appear exceedingly simple, this was in fact the first time an infectious organism had definitively been demonstrated to cause a disease. It was the knockout blow to Galen's miasma theory of disease, and the debate about the cause of many diseases shifted to infections.

After finishing his work on anthrax, Dr. Koch received an appointment at the Imperial Health Bureau in Berlin, where for the first time he was afforded proper laboratory space and research assistants. During the years 1880 and 1881, he continued to lay the groundwork for the modern study of infectious diseases. He developed new ways of growing bacteria, experimenting with different culture media, such as potato and a condensed algae protein called agar. He used steam and chemicals to promote or inhibit growth of bacteria within his new system. He also built on the technology of the microscope, using oil immersion to improve magnification, and employing condensers and different lighting conditions to improve resolution. Dr. Koch was the first to take photographs of bacteria, showing the world the hidden ecosphere that operated beneath the surface of what was visible. Within his laboratory he ignited the golden age of bacteriology, and with his findings he helped advance the understanding of the infectious basis of many diseases.

In August 1881, Dr. Koch traveled to London to attend the International Medical Congress with many of the leading scientists of the day. There he demonstrated his recent advances in bacteriological techniques, earning praise even from his rival infectious-disease scientist Louis Pasteur, who proclaimed, "*C'est un grand progress, Monsieur!*"[21] When not presenting his work, Dr. Koch listened to lectures on different diseases, among them TB—a popular subject given its ubiquitous nature. Dr. Koch left London determined to apply his knowledge to identifying its cause.

A mere eight months later, on March 24, 1882, in a lecture to the Berlin Physiological Society, Dr. Koch presented his work of the previous few months, changing the field of TB forever. Thirty-six scientists were present, many of them well established in their careers and highly respected. But the lecture concluded with stunned silence, not even a whisper of a question or a hint of applause. They had just witnessed medical history and had recognized it as such. One of those present was future Nobel Prize winner Paul Ehrlich, who later stated: "The evening stands in my memory as my greatest scientific experience."[22]

The lecture given by Dr. Koch is held in such high historical esteem not just because of the revelation that he had discovered the cause of the most lethal infectious disease in the history of mankind, but also because of his method of delivering that revelation. Starting off slowly, Dr. Koch meticulously explained the staining techniques he had used to finally culture this elusive bacterium, sharing his observation that older dyes, from the previous decade, worked better than newer dyes because the older ones contained ammonia, which TB cells liked to use as a building block to make their cell wall.

But what gave Dr. Koch's lecture legendary status was what he brought with him. He transported his entire laboratory to the lecture room, including microscopes, test tubes, and slides with bacteria. He showed the audience his dissections from guinea pigs, which he had variably infected with TB from apes, humans, and cattle. The pathology of the different guinea pig lungs was the same, as were the cultures. He announced that everybody present was free to analyze his work with their own eyes, and to repeat his experiments in their own laboratories. Astonished, the scientists slowly came to the front to look in his microscopes and inspect his dissected guinea pig tissue. This was true open-access research, and it left an impression on those present. For the first time, people had proof that TB was caused by a bacterium, and it was right there for them to look at under a microscope—a rod shaped bacterium two to four micrometers in length.

Word of Dr. Koch's discovery spread quickly through Europe and North America. On April 10, 1882, his lecture was published in the *Berlin Medical Weekly*, and from there news spread through the mainstream press, first published on April 23 in the London *Times*, then on April 24 in the Philadelphia *Public Ledger*, and featured on May 7 in the Sunday edition of the *New York Times*. The ghost murderer that had haunted humanity for thousands of years had finally been identified.

With its dense population, and as the destination of millions of immigrants in the past and present, New York City has historically been on the front line of the TB epidemic in the United States. This was never more true than in the late nineteenth century, when TB was the leading killer in the city, claiming ten thousand lives every year, or some twenty-seven patients on average every day.

At the end of the century, Dr. Herman Biggs changed the course of the infection and the fortunes of a city like few in history. Biggs was on the staff at Bellevue Hospital and also worked at the city's Department of Health. Here he introduced new measures to control the rampant TB epidemic, many of which are still in place. However, he was confronted by a medical establishment that didn't like being told what to do and how to practice medicine, challenges that persist today.

Dr. Biggs firmly believed in Dr. Koch's germ theory, and that belief guided his first recommendations, which were also some of his most controversial proposals. In an effort to track the disease and ensure that patients received the most current care and resources, Biggs wanted all cases of TB to be reported to the city's public health department. This divulging of patient information to the government was a radical idea and threw the medical establishment into a state of apoplexy. And as if this weren't enough to provoke outrage, Biggs also wanted the public health authorities to track down all of the patients' contacts, further infuriating the medical establishment.

Physicians at the New York Academy of Medicine swiftly organized against Dr. Biggs, calling his recommendations "mistaken, untimely, irrational and unwise," and further labeling his measures "offensively dictatorial." Citing the importance of patient-doctor confidentiality, they united against the "aggressive tyrannies of the Health Board." The case went all the way to the New York state senate, with Dr. Biggs eventually convincing lawmakers that such measures were necessary for any attempt at controlling TB to be successful. In subsequent years, only about half of New York City's doctors followed the recommendations, and Dr. Biggs didn't push it further, but through compromise he established an important precedent of infection control.[23]

Other parts of Dr. Biggs's program were less controversial. He set up a system in which all patients, in public and private hospitals, could obtain, free of charge, sputum analysis from the laboratory of the Department of Health to determine whether TB was present. He advocated for nutrition and rest, educated patients and their families to cover their mouths when they coughed, and instructed them to dispose of sputum in a sterile way. He also helped establish tuberculosis wards in hospitals throughout the city, where infected patients could be isolated from the general public.

Apart from medical initiatives, Dr. Biggs introduced other pioneering public health measures. The health department issued circulars educating the public about TB, which were translated into German, Hebrew, and Italian to accommodate the burgeoning immigrant population. In 1902, the Committee on the Prevention of Tuberculosis was founded, and the group expounded upon the importance of hygiene through public exhibits and parades. Together, these efforts represented the first mass education campaign against a single disease, like the campaigns we are familiar with against diseases such as HIV and Ebola, and later COVID-19.

The efforts of Hermann Biggs and others at the health department went a long way toward improving the rates of infection and death from TB in New York City. In 1900, there were 280 deaths per 100,000 people per year. By 1920, deaths were down to 126 per 100,000, and by 1940 the number was 49, less than 25 percent of the 1900 level.[24] This was all without the use of any antibiotics. If nothing else, Dr. Biggs had proved that knowledge and prevention could be even more powerful than pharmaceuticals in controlling the spread of disease, especially an infectious one.

Dr. Biggs was not just a physician and epidemiologist; he also knew how to work with the political forces in power, squeezing money out of state budgets even in lean times. He worked with corrupt Tammany Hall and reformist mayors alike. He convinced them all that the improvement of society would pay for itself, and that ultimately the public got the health it paid for: "Public health is purchasable. Within natural limitations, a community can determine its own death rate," said Biggs.[25]

In time, though, the lessons from Hermann Biggs were forgotten, and the ancient nemesis again gained a foothold in America's most densely populated metropolis.

TB grows slowly. Compared to a typical bacterium such as streptococcus, which divides every thirty minutes, TB's doubling time is sixteen to twenty hours. Cultures for normal bacteria are usually held for a few days. For tuberculosis, because it grows so slowly, one has to wait eight weeks for a definitive answer on a culture.

If patience is required when growing the bacteria in the laboratory, then patience is also the rule for killing them in the body. Today, a typical course of treatment for a bacterial pneumonia from streptococcus is five to seven days, usually with a single drug. For tuberculosis, the typical course is six to nine months, with multiple drugs being the standard. Recent attempts at shortening this time to even four months have failed.[26] That's a lot of time for something to go wrong, like a patient stopping their drugs prematurely, or drug resistance developing.

Tuberculosis also has a unique life cycle within our lungs, and a unique relationship with our immune system. When it is inhaled, it normally settles in the bottom portion of the lung. The main inflammatory white blood cell that is initially responsible for containing tuberculosis is the macrophage, known in the immunology world as the "big eater," a large glob-like cell that scavenges pathogens such as TB, as well as inorganic chemical debris and even cancer cells. Unfortunately, the macrophage is not the most effective killer, especially in those who have a weakened immune system, such as somebody with HIV infection. For a few, this initial stage of infection can become significantly worse, and spill out of control. Without proper antibiotics, respiratory failure and death ensue.

Fortunately, this kind of rapid lethal progression occurs only in a small minority of patients. But what is strange about TB is that it has the ability to cause a minor initial infection that can be contained, but not completely killed, by the macrophage and then go dormant for years, or even decades. Tuberculosis goes into hiding in some patients, probably deep in the lymph nodes of the chest. It is likely transported to the lymph nodes by the macrophage, which ingest TB but is then unable to completely kill it. When a person gets older, or their immune system weakens for some other reason, TB can spring to life, burst through the macrophage, and cause a new and virulent infection. This is why some patients are offered prophylaxis, preventive medicines to kill any bacteria remaining in the lung or the lymph nodes. Again, patience is required, with nine months being the traditional standard treatment length necessary to eradicate whatever quiescent bacteria remain.

Treatment regimens for TB generally start with four drugs. If the bacteria have no resistance, the number of drugs can be tailored down to two after a few months. This protocol was created in response to data showing that TB has the ability to escape the reach of a single drug by developing resistance. Resistance was noted with use of the first drug for TB, streptomycin, which became available in 1945. Physicians observed that after an initial improvement, some TB patients would then get much worse. Drug-resistant TB had emerged. Even more ominously, multi-drug-resistant TB would develop in the 1950s, followed in 2006 by extensively drug-resistant TB. More recently, totally drug-resistant TB has surfaced, meaning none of the six or seven or eight drugs tested are likely to kill it.

The way TB develops resistance is unique. Typical bacteria, such as streptococcus or staphylococcus, engage in complex warfare with antibiotics and our immune system. In reaction to an antibiotic, these bacteria may develop pumps to actively drive the drug out of themselves. Or they change the makeup of their cell wall so antibiotics can no longer attach themselves. Some bacteria even develop a trap to ensnare the antibiotic once it comes inside its cell. Amazingly, the bacteria even "talk" to one another, trading pieces of their DNA that encode the pumps and traps and new cell-wall ingredients.

None of this subterfuge goes on in TB, no active attempt to evade drugs or trade genetic material or build new cell-wall components. TB bacteria become resistant through spontaneous mutations in their DNA, something that happens in all DNA, but at a very low frequency.[27] If one of these mutations in the TB DNA happens to help it fight off our antibiotics, then the organism with that mutation gets selected out and thrives. The end effect is that drug-resistant TB is the result of random changes in DNA. It has been estimated that if a susceptible TB specimen is initially treated with two drugs, and those drugs are taken appropriately, there is almost no chance the bacteria can emerge as resistant. They simply do not have the ability to mount a complex defense against antibiotics, or to rapidly make significant changes to their physical makeup. But they do have the ability to capitalize on sloppy use of antibiotics. If antibiotic use is erratic, or if the course of antibiotics is too short, spontaneously mutated bacteria have the opportunity to become dominant and multiply, creating resistance. The conclusion on TB resistance is obvious—it is something we have created with carelessness.

*   *   *

With the public health efforts of Dr. Biggs and others, and the introduction of effective antibiotics in the 1950s and 1960s, the rates of TB
throughout the country steadily declined, and TB was declared to be
largely defeated, no longer a threat. New York City would prove that
declaration of victory was premature. In 1980, the rate of tuberculosis
was about twenty-one cases per hundred thousand people. By 1990, the
number of cases in New York City had more than doubled, to almost
fifty per hundred thousand, with a steep upward curve.[28] Even before the
data came out, this was an epidemic doctors in the city had recognized
for several years. People had been coming into the hospital and dying
of tuberculosis at rates not seen in decades.

As the TB data made its way into the newspapers, it triggered hys-
teria and misinformation about causes of the upswing. The early and easy
explanation for the uptick in tuberculosis cases was the HIV epidemic—
HIV causes a weakening of the immune system, thus allowing TB to multiply unchecked. Two doctors, Karen Brudney and Jay Dobkin, rejected
this knee-jerk assessment. As physicians at Columbia-Presbyterian Medical Center, they were located at ground zero for the tuberculosis crisis
and were uniquely situated to draw their own conclusions.

They worked, much as the organism they were tracking, in a slow
and deliberate way, first confirming that a problem existed. They started
by reviewing data from the year 1969, then looked at rates of TB over
the following twenty years. This period had seen a dramatic increase,
with central Harlem at the epicenter, showing a rate twenty times the
national average.

The two infectious-disease physicians then made some general
observations about what was going on in Harlem at the time, beyond
the obvious HIV epidemic, which had started in the early 1980s. They
noticed that in the previous ten years, the scourge of homelessness had
hit Harlem particularly hard, with many people shepherded into over-
crowded shelters. In the hospital, these homeless patients told doctors
that the shelters were a breeding ground for TB.

With hints that something in society, other than HIV, was happening to cause the uptick in TB cases, Dr. Brudney and Dr. Dobkin

started their study. The methods they used were unpretentious. Between January 1 and September 30, 1988, they interviewed all patients with confirmed TB at Harlem Hospital, and found out what was going on in their lives. They asked them about their housing situation, their current address, who paid the rent, and if there was heat or hot water. They asked the patients if they worked, if they were alcoholics, if they were drug abusers, how many sexual partners they had, and if they had a history of blood transfusions or a prior history of tuberculosis. The two physicians got to know their patients, like super-sleuths.

All the TB patients were also tested for HIV, and then at discharge given an appointment at the Harlem Hospital chest clinic, which was modeled after the Bellevue chest clinic service Dr. Biggs had started almost a hundred years earlier. If the patients were HIV positive, they also got an appointment with an infectious-disease doctor. The study then tracked who showed up at their appointments, or if a hospital admission had gotten in the way. After nine months of studying their subjects, the doctors tallied their data.

In total, 224 cases of TB were diagnosed in that period at Harlem Hospital—an astounding number in an American hospital by today's standards. (By comparison, at the hospital where I work in Philadelphia, with a similar population and hospital size, we average one or two cases per year.) In the end, what the team saw in this population was not surprising to them, and it drove home Biggs's observation that a city gets the public health it pays for.

Of the 224 cases, almost 80 percent of the TB-sufferers were male, and about half were alcoholics. About 70 percent were either homeless or in unstable housing. A quarter had been diagnosed with TB previously, and almost all of these patients admitted to not finishing treatment. Of the initial 224, 178 were able to be discharged, with the rest dying in the hospital. Of those making it to discharge, 89 percent did not complete therapy, with most never making it to a single outpatient visit. Although a majority of patients did in fact test positive for HIV, this clearly wasn't the immediate cause of their unsuccessful TB treatment. What had done the most harm was a breakdown in social infrastructure and a lack of appropriate follow-up for a treatable disease.

Dr. Brudney and Dr. Dobkin dug deeper into the causes of this breakdown in TB control. Just as TB doesn't grow overnight, the policies and cutbacks that allowed it to flourish again also did not happen in a day. In 1968, a task force appointed by Mayor John Lindsay reported on the state of TB and future directions. At the time, the city spent $40 million per year on TB, including on clinics and on one thousand inpatient beds. The 1968 task force recommended closing some inpatient beds but continuing strong financial support for outpatient clinics and increased home visits by nurses and home health aides.

Instead, with New York City's fiscal crisis of the 1970s, almost all of the one thousand TB inpatient beds were eliminated, and the budget shrank to less than $25 million in the span of ten years. Financial support at the federal level also shrank, from a high of $1.4 million per year in 1974 to $283,000 in 1980. The recommended increase in home visits never materialized, and opportunities to screen for TB at drug clinics were missed. By 1979, New York City was seeing an increase in its TB cases, even before HIV arose.

Dr. Brudney and Dr. Dobkin published their work in 1991 in the *American Review of Respiratory Disease*. Titled "Resurgent Tuberculosis in New York City: HIV, Homelessness, and the Decline of Tuberculosis Control Programs," their paper succinctly described how a volatile mix of homelessness, drug use, and alcoholism, along with a loss of funding, had gotten in the way of patients being treated appropriately.[29] Tuberculosis, ever the opportunist, took advantage. In a time when we had our most powerful antibiotics, New York was doing worse than Dr. Biggs had done ninety years before, with education and no antibiotics at all.

Their paper was harsh in many ways, but it was also filled with hope in the form of opportunities for change. The resurgence was clearly not just the product of HIV, or some other mysterious affliction like Galen's miasma of bad humors from poor sanitation. People were just not taking their medications, for a variety of personal and societal reasons. They had priorities other than their health, like paying the rent, or finding shelter, or looking for food, or obtaining drugs. Toward the end of their paper, Brudney and Dobkin suggested a relatively straightforward, time-tested approach of getting the epidemic under control: longer inpatient hospital stays, residential TB treatment facilities, and increased and aggressive community-based surveillance of medication administration.

Looking back, we know this paper represented a turning point in the fight against this new onslaught of an old disease. People clearly listened to the advice of Drs. Brudney and Dobkin, and of others on the front line of the new epidemic. Funding increased, as did directly observed therapy (DOT) to ensure people were taking their medications. The stories of creativity and persistence on the part of caseworkers are inspiring in their display of dedication. Caseworkers did DOT at health clinics and in homes, and also at more unusual places, like McDonald's, and even under bridges where homeless people gathered.[30] With an influx of money and motivated people, the rate of TB slowly but surely started to decline. As of 2016, the rate of TB in New York City had reached an all-time low of 6.9 cases per 100,000, down from 52 in 1992. Pockets of increased rates still exist, but nothing like the rate or absolute numbers seen in the late 1980s.

New York City remains on guard, especially as a slight uptick was reported in 2017 to 7.5 cases per 100,000 people, or from 565 total cases in 2016 to 613. Fortunately, this number came back down in 2018 to 559 cases.[31] Newer surveillance methods utilizing unique technologies are helping. Remote observed therapy is now available, with patients logging onto their computers daily and taking their medicines on camera. The genetics of each TB bacterium can be tracked. With the help of DNA analysis, a 2013 outbreak in Sunset Park was localized to Chinese immigrants, with the strain probably imported from China. A small Internet café and a karaoke bar were the likely sites of transmission, and this information was used to find people asymptomatically infected, with the goal being to treat them and thus stop the spread.[32]

*   *   *

The strength and power of tuberculosis in this country are not what they once were. Each year since the early 1990s has seen a decline in the number of cases, and as of 2017, the overall rate of the disease was 2.8 per 100,000 people. The total number of deaths from TB in America was 528 in 2016, way down from 1,705 in 1992. Importantly, there has been no drop-off in awareness as occurred in the 1970s, and completion of therapy remains very high.

But if TB's imprint on us here in the United States is becoming a whisper of what it once was, the numbers abroad tell a different story. The disease is raging out of control in many countries. Absolute numbers of cases are staggering in countries like South Africa, where in 2019 the incidence was 520 per 100,000; in the Philippines it was 554.[33] Resistance has also become more of an issue, with multi-drug resistance rampant in countries like Kazakhstan, Ukraine, and Russia. The first cases of extremely drug-resistant strains of TB were reported in places like Iran, India, and Italy, but cases have since been reported in the United States as well.[34]

Our air today, much like our economy, is becoming more communal and globalized. Drug-resistant infections are a societal problem. Of all the cases of TB in the United States in 2018, 70 percent occurred in non-US-born people, a statistic that is increasing every year. Simply closing our borders to all immigrants is not a realistic solution, but we do screen immigrants and refugees for tuberculosis, for their own health as well as for the health of the country. We also should assume a greater role in helping control the multitude of TB epidemics throughout the globe.

Given what we know from Drs. Brudney, Dobkin, and Biggs, TB is a disease that can be managed. But until we follow through on all the measures at our disposal, the old torment that joined us on the plains of East Africa as we emerged as a species is going to continue to haunt us. It's not as smart as we are, but it has one thing that we often lack—patience.

* * *

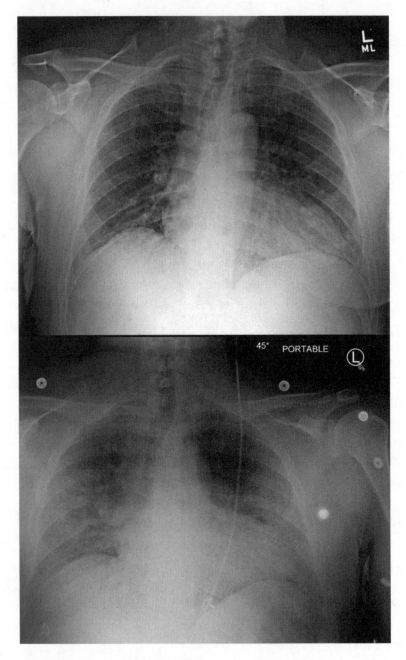

X-ray above with mild COVID-19 pneumonia at the bases, and X-ray below one day later with dramatic progression.

Herman Biggs's observations about the costs and benefits of public health never felt more relevant than in 2020, when the world literally shut down to try and stop the spread of the novel COVID-19 virus. It is clear from early analysis that many missteps were made, both in the United States and in countries throughout the world. We found ourselves in a situation in which a respiratory virus brought the world to its knees, with everybody scared about contracting a lung infection.

Coronaviruses do not historically cause significant disease in humans. A virus that typically infects birds, cats, pigs, and bats, it started causing trouble only when humans came into close contact with animals that harbored a rare, mutated version of the virus that could infect us. The first of these catastrophic events transpired in South China in 2002, when a coronavirus in civet cats jumped to humans.[35] The close contact and slaughtering that takes place at exotic meat markets was the likely conduit of transmission. A home video of these markets shows meat from exotic animals stored in less than sanitary conditions, with no refrigeration or even ice.[36] Despite the first SARS coronavirus sweeping through Guangdong province and then spreading globally in 2003, infecting eight thousand people and killing eight hundred, an ineffective effort was made on the part of the Chinese government to close unsanitary exotic meat markets.

Rumblings about a new coronavirus began to be heard in the United States in January 2020, and it soon became clear that a large outbreak in China had been building throughout December. The origin was again an exotic food market, this time in the city of Wuhan, in the Hubei province. The likely source was either a bat, a popular commodity at these markets, or possibly a pangolin, a small scaly anteater native to Asia and also available at open-air meat markets.[37] Like the first SARS virus, COVID-19's new preferred habitat was the human respiratory system, and it spread when somebody coughed and expectorated the virus into the air, or one of the droplets settled on a surface and was subsequently touched by another person.

In January and February 2020, leadership in the United States and at the CDC took opportunities to downplay the threat, then reinforced

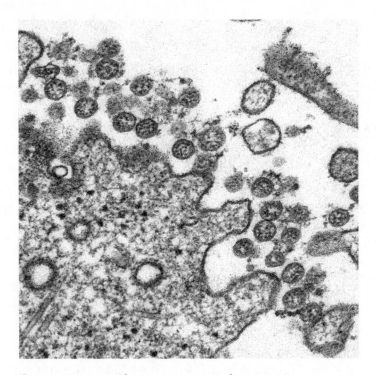

Coronavirus particles, as seen on an electron microscope, attempting to infect a cell.

this careless response by neglecting to make preparations or take precautions, such as improving testing facilities or restricting travel. Dr. Helen Chu, at the University of Washington, was one of the people who, in January 2020, wanted to investigate the possibility of the novel coronavirus already circulating. She had been collecting nasal swabs locally in Seattle for an unrelated viral project, and she asked the state and federal governments for permission to test the swabs for coronavirus. Both said no. Growing nervous about an outbreak, Dr. Chu and her group went ahead and tested the samples anyway—and found a positive result in somebody with no known travel out of the country. The virus was circulating without anybody's knowledge, or apparent concern.[38]

Although not all of the data is available, one key to a country controlling this disease appears to be fast and easily available testing. This would let people infected self-quarantine, while allowing for screening for those who had close contact with them. These people could then self-quarantine if their test was positive, slowing the spread of disease significantly. South Korea implemented this approach with apparent success, even using cell phone technology to track where infected people had been to appropriately sound the alarm. The South Korean government also quickly ramped up the ability to test ten thousand people a day, often at convenient drive-through testing stations that minimized potential exposures.[39]

In the United States, testing was available from state laboratories, but their technology was often woefully antiquated, and up until mid-March 2020, state lab results in Pennsylvania would take four or five days to come back. Even more concerning, up until the second week of March, the state of Pennsylvania could run only five or six tests a day because of limited equipment. Private labs such as LabCorp and Quest stepped in to help fill the void, but only after a month had been squandered. And even then, the private labs understandably did not want patients coming directly to them for specimen collection, leaving potentially infected people with no choice but to put on a mask and wait at the emergency room. Fortunately, rapid drive-through testing stations were eventually implemented.

The coronavirus story is one unfortunate illustration of how our air is communal, that the world is interconnected as never before, and warnings about potential global health threats need to be taken very seriously.

Chapter 7

Nicotine Seduction
and Stem Cells

As the site of gas exchange for the entire body, the lungs must maximize air flow. However, when we hijack this efficient system for other purposes—such as inhaling substances that bring us pleasure—bad things can happen. We've been exploiting our gas-exchange system for thousands of years, but only in the last few hundred have we taken it to such an extreme, and to our own detriment. Specifically, through the widespread use of tobacco, some have learned to stimulate their brains in a unique way that has changed the entire field of pulmonary medicine.

For millennia we have used the lungs, through smoking, to transport substances. This should not be surprising, since the lungs provide one of the fastest routes to deliver drugs to our brain. There is something calming and relaxing about smoking as it is part aromatherapy and part calming medicine, and its role in our cultural history has been enormous.

Smoke has been utilized throughout history by almost every culture and religion. Egyptians, Babylonians, and Hindus all burned incense as an offering to their gods. Historical evidence of direct inhalation of smoke is also abundant, starting in ancient China with the use of cannabis, which spread to India, the Middle East, and Africa. In the fifth century BCE, the Roman historian Herodotus documented how Scythians, an ancient nomadic people of what is now southern Siberia, placed hemp seeds on hot stones, "and immediately a more agreeable vapor is emitted than from the incense burnt in Greece. The Company, extremely transported with the scent, howl aloud."[1]

Just as the Scythians did several thousand years ago, our culture has taken advantage of the efficiency of the lungs to transport drugs. We have unfortunately taken the utility of smoking completely out of balance. Respiratory illnesses in this country have exploded over the past fifty years, increasing 163 percent from 1965 to 1998. The increase has continued more recently, though at a somewhat slower rate, with a 30 percent increase in the death rate from respiratory illnesses between 1980 and 2014.[2] Mortality rates associated with lung disease have stabilized since 2014, but the continued burden of respiratory illnesses is undermining the narrative in this country that we are continually getting healthier and living longer than the generation before. Widespread use of tobacco is a huge part of the problem.

Lung physicians know these diseases well and see firsthand every day what they can do to people's lives. To combat the tobacco scourge, most lung doctors take time at each office visit to counsel patients on quitting smoking. A few have focused their careers on the problem, and from them lessons are available that will help us bring this scourge under control.

Mr. Johnson was frustrated the first time I saw him in my hospital's pulmonary clinic. At rest he was fine, but whenever he tried to do something that demanded the slightest increase in his metabolic rate, he became short of breath and had to stop. Mr. Johnson could not take in enough air to do what he wanted to do in life.

He was only in his late forties, but he told me he had already been smoking for some thirty years, often more than a pack per day. This was more than enough time to make him part of the unlucky 5 percent who develop very serious lung disease from smoking. I sent him for a breathing test, and fifteen minutes later had the answer to the question of what was causing his breathing problems—when he exhaled, he could not get the air out of his chest because his airways had lost the needed elastic tissue and had been rendered floppy from the years of smoking.

"You have COPD," I told him. "Chronic obstructive pulmonary disease."

"Is that like emphysema?" he asked. I replied yes, it was the same disease.

He looked at me and asked earnestly how we were going to fix this, proclaiming that he was here and ready, that I was the doctor and he trusted me. I didn't say anything right away, trying to let some of the expectation dissipate. I then asked him if he had any other medical problems. He told me that when he was in his teens, he had had a bad case of pneumonia, but it had been cured effectively with antibiotics. I wondered if now he expected me to pull out my prescription pad and write him a script for a few tablets, much as the doctor had done for his pneumonia, curing his lungs of COPD after a few weeks of therapy. Except there were no tablets to give him, no medications beyond inhalers, which would only help him feel a little better.

I asked him if he still smoked, and he replied yes. Still a pack a day. After another pause, he continued, stating he was discouraged with his breathing and his life, and he was here to get better. I kept quiet, trying to think of the best way through this. When Mr. Johnson finished, I stayed silent for a moment more, and then started talking slowly and carefully. "You're frustrated because your world is small right now," I said to him, making a small box with my hands. I was pushing him emotionally to acknowledge what was going on. "You can't do the things you want to do because of your very difficult breathing, and you're angry at yourself, and you're angry that your world is small."

After a long silence he replied. "Yes, yes." He shook his head. "I'm frustrated and angry. I can't do what I want to do."

We went over his breathing tests with brutal honesty. At almost fifty, he had horrible lung function. Many eighty-year-olds who have smoked for a lifetime had better lung function than he did. I prescribed an inhaler to help provide a measure of relief and to hopefully take the edge off his frustration. At the end, I told him what he already knew. "You need to quit smoking."

"I know," he replied, and left.

* * *

When a smoker like Mr. Johnson lights a cigarette and inhales, the smoke rushes past the vocal cords, through the trachea, into the bronchi and bronchioles, and finally into the alveoli. At that point, the smoker pauses, allowing the nicotine a moment to pass through the barrier of the lung tissue and into the capillaries, where it is then carried to the brain. The remaining smoke gets exhaled, the cloud expelled into the atmosphere.

The years of toxic inhalation does different things in different parts of the lung. One of the first effects it produces is inflammation at the level of the bronchi and bronchioles, where normal mucus-secreting cells, called goblet cells, begin producing copious amounts of mucus in reaction to the irritation. This is why many smokers carry a diagnosis of chronic bronchitis (inflammation of the bronchi) and have a hacking cough that produces yellow and green sputum.

With ongoing smoking, the next change that begins to happen is cell death. Cellular death is normal for all organs of the body, and the lungs are no different: every day, hundreds of lung cells die and are replaced. The problem with smoking is that the death of cells is accelerated, while the replacement of cells slows down. In the bronchi and bronchioles, it is believed that the unchecked death of the airway basal cells is particularly problematic. The airway basal cell is a short, squat, cuboidal cell that lies deep within the tissue of the airway. Although it is not present in overwhelming numbers, its job is critical as it functions as the stem cell for other airway cells—differentiating and then multiplying into the other workhorse cells of the airway, like the squamous cell.

As squamous cells and others die off and fail to get replenished, the airway loses its tone and gets floppy. This is not such a big problem for when we inhale, but during exhalation these floppy airways collapse, and air gets trapped in the lung along with carbon dioxide. This is why some COPD patients have a barrel-shaped chest from trapped air. Others tend to develop thin lips as they constantly try to slow down the flow of air to avoid collapse of the bronchioles and bronchi. Ventilation becomes a challenge, and carbon dioxide, and subsequently acid, begins building up in the blood.

While the damage in the airways progresses from inflammation to destruction, deeper down in the alveoli it is usually simply destruction that occurs. The cells that comprise the gas exchange units of our lungs, type I alveolar cells, begin dying off. The source of replenishment for these cells, the type II alveolar cells, also begin to die off. Vast holes develop in the lung, empty spaces of air where no gas exchange is occurring, like a sponge whose holes have tripled in size. Now, getting oxygen into the blood becomes a problem as well.

When I saw Mr. Johnson back in the office a few months after the first visit, I immediately noticed something different about him. His eyes appeared clearer, his skin more radiant, his hair softer and neater. He appeared more alert, more alive, and the stale smell about him was gone. I had seen this transformation several times before and asked him right away the question I thought I knew the answer to: "Did you quit smoking?"

He answered without hesitation: "Yes, Doc, I got rid of those cigarettes. I haven't had a cigarette in over a month."

We took another look at his lung function. The typical total amount of air in the chest for a man his size is a little more than four liters. With the years of smoking and the damaged airways, Mr. Johnson had over five liters in his lungs—the absence of tissue meant more air was filling his chest cavity, and with floppy airways he couldn't inhale and exhale at the normal speed. Rather than being able to exhale all of his air in two seconds, a typical amount of time for healthy lungs, he needed eight seconds. Given that our ability to move depends on energy that comes from the flow of air through our lungs, this reduced air flow slowed Mr. Johnson's world down by a factor of four.

We did what we could, and over the months and years we worked together to get the most out of what remained of his lungs. I stressed to him, as I stress to all patients with COPD, that he didn't get sick in a day, and he would not feel better in a day. The same time and effort and money he put into cigarettes and smoking he now needed to put into getting his body back in shape. He promised me he would, and true to

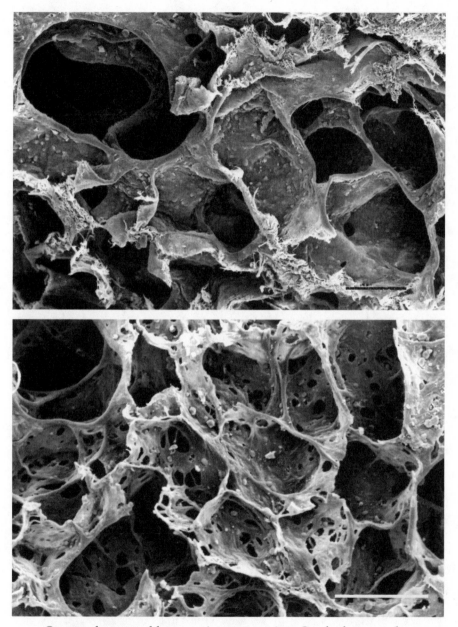

On top, the normal lung cut in cross section. On the bottom, the
lung of a COPD patient with prominent holes.

his word, he stayed off cigarettes, used his inhalers, and went to physical therapy to get stronger. With the additive effect of a few small things, his life improved tremendously, and even with severely diseased lungs he is getting vastly more out of life today than when I first met him.

The tobacco plant is indigenous to the Americas and wasn't known in Europe, Africa, or Asia until explorers returned with it in the sixteenth century. Before then, Native American civilizations had used tobacco ceremoniously and as a gesture of good will at the end of bartering to seal a transaction.

One of the first descriptions of tobacco in Western literature came from the Spanish historian Bartolomé de las Casas, who in 1527 noted that the indigenous people in Cuba used "certain herbs to take their smokes, which are some dry herbs put in a certain leaf. [They] receive that smoke inside with the breath, by which they become benumbed and almost drunk, and so it is said they do not feel fatigue." De las Casas also was aware of the addictive power of tobacco, observing: "I knew Spaniards on this island of Española who were accustomed to take it, and being reprimanded for it, by telling them it was a vice, they replied they were unable to cease using it. I do not know what relish or benefit they found in it."[3]

A few decades later, tobacco as a cash crop took off with the establishment of the Virginia colony, where it was mass-produced and exported by English settlers. Sir Walter Raleigh is credited with first bringing Virginia tobacco to Europe in 1578, where its health effects were lauded. In 1595, the Englishman Anthony Chute published the pamphlet *Tabaco*, outlining how the plant had therapeutic potential if rubbed on the skin or inhaled, and how it relieved any sense of fatigue at the end of the day. His printer, Adam Islip, added that it "cured a gentleman that long languished of a consumption," and also wrote somewhat ironically that tobacco "hath many strange vertues which are yet unknowne."[4]

Not everybody was receptive to the tobacco plant. The British king James I, in 1607, wrote one of the first treatises against tobacco, the use of which he described as "a custome lothsome to the eye, hatefull to the Nose, harmefull to the braine, dangerous to the Lungs, and in the blacke stinking fume thereof, neerest resembling the horrible Stigian smoke of the pit that is bottomelesse."[5]

Whatever the feelings of the king may have been, the economic reality was that tobacco drove the early economies of the Virginia and Carolina colonies. The English settler John Rolfe was one of the first to benefit financially from its mass production, introducing the plant *Nicotiana tabacum*, a native of Bermuda, into Virginia. In 1620, forty thousand pounds of tobacco were exported to England. Its use and economic importance grew, and during the first one hundred years of American colonialism, excise taxes from tobacco made up a third of the revenue collected by the British government.

Early use of tobacco centered on pipes, chewing tobacco, snuff, and cigars. In the American South, it was a ubiquitous drug, as journalist Sidney Andrews observed in 1866 during a tour of North Carolina: "The amount of tobacco consumed by the people is beyond all calculation. At least seven-tenths of all persons above the age of twelve years use it in some form. Nearly every man and boy smokes or chews, and very many of them do both, while the country women chew and smoke to some extent, and women of most classes 'dip.'"[6] But even with this level of consumption, profits stagnated after the Civil War. The tobacco industry searched for something new to boost revenue, and one person found it in the mass-produced rolled cigarette.

James Buchanan Duke, known as Buck Duke, was born in 1856 near Durham, North Carolina, and at age twenty-four was managing a factory making hand-rolled cigarettes. At the time, the rolled cigarette was a niche product, with most Americans preferring pipes, chewing tobacco, or cigars. His factory was modest, with each worker able to produce about two hundred rolled cigarettes per day. Not happy with this output, Buck Duke thought a new invention, by James Bonsack, could improve it.[7]

Others had also looked to increase the efficiency of producing the rolled cigarette, and in 1875 the tobacco company Allen & Ginter, of Richmond, Virginia, offered a reward of $75,000 to anyone who could invent a machine that could roll cigarettes. Eighteen-year-old college student James Bonsack decided to drop out of school and dedicate himself to winning the prize. After several years of work, by 1880 he finally had his invention, his patent, and the reward. His machine was a major upgrade from the hand-rolled factory method and was capable of producing a remarkable 120,000 cigarettes per day.

Although impressive in its promise to mass-produce cigarettes, the machine suffered from mechanical breakdowns. Also, the now-open ends of the cigarette tended to dry out, whereas the traditional hand-rolled cigarette had twirled paper at either end. All the major tobacco companies declined to invest in the new machine, but Buck Duke struck a deal to work with Bonsack to improve his device. Duke labored with his own engineers to iron out the mechanical issues, and they applied additives to the ends of the cigarette—glycerin, sugar, molasses, and a few chemicals—to prevent them from drying out.

Their efforts paid off, and Buck Duke's machine started successfully cranking out more than one hundred thousand cigarettes per day. Since he had no market for these cigarettes, Duke created one, giving away his cigarettes at beauty pageants and other events. He took out ads in the new "glossies," the first magazines in the United States. He put baseball cards in the packages to capitalize on the rising popularity of the sport. In 1889 alone, Duke spent the equivalent of $25 million on advertising. His new cigarette also fit the changing lifestyle of Americans: rolled cigarettes were suited to the increasing urban crowd, with no messy spit to dispose of or dirty pipe to carry around. They were very portable, had a modern look, and lit quickly, making them easier to use during coffee breaks and in restaurants.

Buck Duke's aggressive advertising paid off. Acquiring money and market share, he bought out four of his competitors and in 1890 merged them into the American Tobacco Company. He also saw potential overseas and formed the British American Tobacco Company to expand into the European market, taking advantage of his one-size-fits-all product, easily recognized and branded as *Duke Cigarettes*.

Duke became extraordinarily rich and in 1924 gave away some of his money to the then-small Trinity College in Durham, North Carolina. In return for his gift of $100 million, the leaders of the college honored Duke by renaming the institution after him. Ironically, his donation amounted to $1 for each of the estimated one hundred million lives lost to cigarette smoking during the twentieth century.[8]

Though certainly a master businessman, advertiser, and human psychologist, Buck Duke had something else on his side that made his job easy. He had one of the most addictive drugs known to humankind, and he gave it a delivery system that transported it to the brain in seconds. The rest was just Madison Avenue spin.

At the beginning of the twentieth century, Buck Duke used all his energy to get people to start smoking. Some hundred years later, Frank Leone is trying just as hard to get people to *quit* smoking. And because smoking has such a deleterious effect on people's lives—it affects lung health, heart health, and brain health, increases the risk of many lethal cancers, negatively affects mood, exacerbates depression, impairs sleep quality, and puts children's lives at risk—Leone is actually doing much more than helping people quit smoking. Through a deep understanding of neurobiology, he is giving people back their free will, their ability to choose, and thus their entire lives.

Leone talks to smokers looking to quit with the passion of a preacher. He grew up in New York City and later attended the University of Pittsburgh and its medical school. He moved to Philadelphia in 1990 and did his residency at Thomas Jefferson University Hospital, where he saw that Philadelphia had a serious tobacco problem, with a smoking rate of 28 percent of the population, more than ten percentage points above the national average.

But what motivated Frank Leone to dedicate his life to getting people off cigarettes wasn't the statistics. Lung doctors know the majority of diseases they see are somehow linked to cigarettes. As a young lung doctor, Leone would frequently try to get patients to stop smoking, and like most physicians, he didn't have much luck. Nonetheless, he considered his patients' inability to stop smoking as their problem, their weakness, their inadequacy. Eventually, however, he realized he had to own his patients' conditions and addictions and come up with a plan for improvement, just as he did with any other pulmonary disease.

Frank Leone also recognized that in order to solve the mystery of cigarette addiction, he needed to understand the neuroscience and the psychology of tobacco addiction. He needed this knowledge because he has one of the most difficult jobs in the world. Quit rates are abysmal; if a tobacco cessation product has a quit rate of 20 percent, that's considered a huge success.

One question Leone asks his smoking cessation classes regularly is, "Why does a bright, intelligent, motivated adult, with a million reasons to stop smoking, who has tried a million times, still find it so difficult to quit?" And he goes on to empathize: "Who hasn't had a friend look at them and say, 'Why don't you just stop?' To the smoker, this sounds like the friend is telling them to grow a third eye."

This is the power of the cigarette, the grip of nicotine. Yet, the addictive nature of smoking is hard to explain. It is unlike that of other drugs. We are aware that when one snorts cocaine, injects heroin, takes a shot of whiskey, or smokes methamphetamine, a high comes along with it. Nicotine offers no such high, no effect of being "stoned," no significant change in sensory input or experience. It is not euphoric or reality-altering in any respect. One doesn't escape the world or drown one's sorrows with cigarettes, which begs the question of where the addictiveness of this drug is coming from.

The answer is found deep in our brainstem, the ancient part of the brain that lies near the base of our skull. Within it lies the ventral tegmental area (VTA), which can be thought of as a relay station for all the danger and survival signals that come into our head.[9] These signals get processed in the VTA and screened for threat level. If a barking dog with bared teeth is coming at us, that signal is sent to the VTA and gets prioritized as something important. The VTA then sends out a signal to another part of the midbrain, the nucleus accumbens, which amplifies the signal and creates motivation to do something, in this case probably to run away from the dog, and fast.

If one can get away from the dog to a safe place, the brain then needs a signal to turn off the VTA, to calm it down so it stops sending out danger signals to the nucleus accumbens. This function is performed by the neurotransmitter acetylcholine, which latches onto the VTA to cool things down. Many of the receptors in the VTA are, not surprisingly, acetylcholine receptors, and there are several different types, some called nicotinic acetylcholine receptors because nicotine can also bind there, creating an imposter safety signal.

So rather than a high, nicotine creates a sense of well-being that all is good and safe in the world. It is a powerful emotion, so powerful that it easily overrides the rational understanding that cigarettes are unhealthy. The drug nicotine is an "invisible hand," guiding smokers to a feeling of calm and well-being. Cigarettes manipulate the mind at a deep neurobiological level, taking away not only our health, but our ability to choose. And it all starts with the lungs as the gateway.

When you light a cigarette, the paper and tobacco explode into a bright orange of combustion and heat. The temperature at the burning end of a smoldering cigarette is an astonishing 900 degrees Fahrenheit, and with a deep drag, oxygen feeds the fire, causing a burst of orange that reaches about 1,200 degrees Fahrenheit.[10] The air that the smoker inhales is a toxic mixture of nanoparticles floating in nitrogen, oxygen, and carbon dioxide, mixed with the deadly gases carbon monoxide and hydrogen cyanide.

The particles in the smoke itself contain nicotine, which has been used as an insecticide, but also approximately seven thousand other ingredients. The gas of tobacco smoke also contains benzene, which is found in rubber cement; cadmium, an ingredient in battery acid; and arsenic, as in rat poison. The list of known toxins and carcinogens includes acetone, toluene, DDT, butane, and naphthalene.[11] So-called natural cigarettes, without some of the additives, are thought to be no safer, as the dried tobacco leaf stands on its own as a toxic product.

This mixture of toxic gas and poisonous nanoparticles makes its way down into the deepest part of the lungs, where it quickly spreads out into the vast network of alveoli. Here, it crosses easily into the blood, flows to the heart, and within ten seconds of inhalation is attaching itself onto acetylcholine receptors in the brain's VTA. There is no more efficient drug delivery system than cigarette smoke via the lungs. Light, inhale, wait a few seconds, and then feel the rush of calm descend.

Nicotine binds very strongly to the VTA, and once acclimated to nicotine, the brain likes to keep its levels constant. The half-life of nicotine is about two hours, so over the course of a day, smokers are in a constant battle to maintain steady levels. In the morning the brain is more desperate, after eight hours without its fix. Heavy smokers often wake up, grab a pack of cigarettes strategically placed next to the bed, and get a quick inhalation before doing anything else.

In addition to creating a short-term sense of well-being, smoking causes other, more permanent changes in the brain, including turning off certain genes. Some of these don't turn back on for years after a smoker has quit. So when smokers try to quit, they are fighting short-term changes in receptor sensitivity and a constant need to maintain nicotine levels, as well as longer-term rewiring and genetic changes. We like to think that quitting tobacco is a one-time event, and once you quit, you're cured. But with these long-term changes, quitting is much more like fighting a chronic disease, one that needs constant attention throughout the rest of one's life.

Considering the neurobiology of nicotine addiction, it's a wonder that anybody can quit smoking. But Frank Leone is convinced it can be done effectively and painlessly, and he helps people do it in a way that is the exact opposite of what most doctors tell them to try. Leone, quite purposely, does not focus on the smoking, but rather on the dependence, the reason people feel they need to use nicotine to create a constant sense of well-being.

He also knows that all smokers who come to him have a conflict within: the thinking part of their brain that wants to quit, while the instinctual part of their brain is not letting them. They have, as he describes it, extreme ambivalence, a deep conflict that is both psychological and biological.

Leone knows that he must be careful with this ambivalence. If he pulls too hard on the part that wants to quit smoking, the boomerang effect of the midbrain—the VTA—pulling back will take effect. His main weapon for fighting this ambivalence is aggressive nicotine replacement, since to him there is no such thing as toughing it out with the brain. A smoker's brain chemistry will win every time. The system that is controlling them, the central portion of which is the ventral tegmental area, is evolutionarily millions of years old, hardwired into one's survival circuitry. A common joke in tobacco cessation classes speaks to the challenge: "Quitting smoking's easy. I've done it a hundred times."

Of the total smokers in the United States, about 70 percent say they want to stop smoking entirely. Of those, about half tried to quit in the past year, but success rates are abysmal. For those who don't use nicotine replacement, only about 7 percent remain tobacco free at six months, while those who use various types of nicotine replacement average quit rates of 15 to 20 percent.[12] Varenicline is a newer medicine that works in the brain to mimic the effects of nicotine. With sales of $755 million per year worldwide, it is a blockbuster drug, but it, too, has low quit rates, also in the 15 to 20 percent range.

Frank Leone knows these statistics by heart. He counsels patients to use multiple options for treatment, all at once if necessary and tolerated. He feels fears of nicotine toxicity in replacement products are overblown, since many people who smoke two packs a day for decades are fine, so extra short-term nicotine is worth it in the service of quitting. The key is to trick the brain into thinking it's still getting its fix, and then slowly wean off the need for the replacement.

The lungs have a remarkable ability to repair themselves, but after smoking goes on for years and COPD develops, much of the lung is destroyed and cannot regenerate. Stem-cell therapy holds enormous promise to fill the holes that have formed and return the lung to a healthy state. And stem cells are just part of a larger plan of organ regeneration that could one day change the field of medicine, and change how we approach injury and repair not just in the lungs but throughout the body.

One of the scientists at the forefront of lung regeneration efforts is Darrell Kotton, head of the Center for Regenerative Medicine at Boston University School of Medicine. Since completing his fellowship in 1997, he has devoted himself to the single question of how to rebuild the lung after an injury. When he began, Dr. Kotton noticed that almost all of the therapies available for lung diseases, including COPD, could at best be said to abide by the time-honored Latin phrase *primum non nocere*, first do no harm. With his work in regenerative medicine, Dr. Kotton wants to shift the paradigm to *primum succurrere*, or first hasten to help. It is a simple but profound change in expectations.

After the egg and the sperm connect, for a moment they exist as a single cell. This quickly begins to divide, and then divide again, until at about day five we have what is called a blastocyst—a tiny circular structure, less than one millimeter in size, that is made up of about two hundred cells. It has a ring of outer cells, called the trophoblast, and a mass of inner cells, aptly called the inner cell mass.

This inner cell mass is extremely important; it is made up of stem cells that are functionally pluripotent—any one of these cells has the potential to develop into any other cell in the body, whether in the lung, the brain, the heart, or the skin. They are completely undifferentiated—it is not yet determined what type of cell they will be.

In the course of development, this inner cell mass of undifferentiated cells lasts only a fleeting few hours. It quickly begins to differentiate into cell lines destined to become different organs, making stem cells in this state almost impossible to study. But a breakthrough occurred in 1981, when Dr. Gail Martin at the University of California, San Francisco, and Drs. Martin Evans and Matthew Kaufman at University College London were able to isolate and maintain the undifferentiated cells for study.[13,14] This breakthrough offered the prospect of being able to reproduce what the human body knows how to do instinctively, namely to assign a pluripotent cell to differentiate into a cell of the organ of our choice.

The key to success in regenerative medicine lies in what scientists in the field call *directed differentiation*. Investigators must decipher the different and complex chemical signals produced within the inner cell mass that guide each cell to its final destination. This daunting task got easier in 2006, when Dr. Shinya Yamanaka of Kyoto University in Japan managed to reprogram skin cells from an adult and turn them into pluripotent cells similar to those in the blastocyst.[15] This earned Dr. Yamanaka the Nobel Prize in medicine and also eliminated the ethical considerations of working with embryos, the original source of stem cells used in research. Just as important, this new process allowed scientists to work with a patient's own cells, making it much easier to introduce them later without fear of rejection. Since 2006, various other cells have also been taken back to a pluripotent state.

Back at Boston University, Dr. Kotton has begun to guide these pluripotent cells toward final differentiation of a lung cell. In 2017 he published a paper in which he specifies the exact factors needed to take somebody's own blood cells, engineer them back to a pluripotent state, and then bring them forward to one of the most common lung cells, the alveolar type 2 epithelial cell (the cell that secretes surfactant and is also the stem cell for the rest of the alveolar cells).[16]

Dr. Kotton also described the process of successfully retooling the cells from the blood of patients with a genetic disease that causes their type 2 epithelial cells to produce defective surfactant. He began by engineering pluripotent cells from the patients' own blood cells. Then, using a cutting-edge technology called CRISPR, he genetically corrected the surfactant defect. Finally, he brought these re-engineered blood cells forward to mature type 2 cells without the genetic defect these patients had lived with their whole lives—an amazing accomplishment.

The next big step in this journey of regenerative medicine will be to devise a procedure for getting these cells back into the patient, a process called engraftment. This may be very difficult in the lung, which, over the course of its evolution, has built up immense immunological defenses. If successful, though, this procedure would complete the circle of regeneration: extracting a blood cell, bringing it to a pluripotent state, modifying its defect, moving it forward to a lung cell, and then placing it back. The range of lung and other diseases that could be improved, even cured, is vast.

Darrell Kotton cautions that efficacious lung regeneration is still many years off. Other organs, such as the eye, will likely see results much sooner than the lung, as engraftment will be more easily accomplished there. Dr. Kotton also feels strongly that his research should be available to all—he has made pluripotent cell lines engineered from different lung diseases available to any researcher in the world, and when he publishes his results, it is almost always in open-access research journals, available to all without cost. He advocates that reciprocity and exchange of ideas among researchers is the best way we can shift from *primum non nocere*— doing no harm—to *primum succurrere*—hastening to help.

The lessons we have learned from a century of widespread tobacco addiction are many, and there are numerous positive signs today, especially in this country, where smoking rates have dropped to a remarkable low of 13.7 percent in 2018.[17] Considering the rate was above 40 percent in the 1960s, this is a huge step forward.

Frank Leone, however, is guarded in his optimism. In the 1950s, addiction to drugs and alcohol was thought to be primarily a question of moral character; then we decided a biological dependence was driving this behavior. In the 1970s and 1980s, addiction came to be seen as a problem involving neurotransmitters and receptors, with the solution being to block these receptors with drugs like methadone in the case of opiate addiction or nicotine replacement for tobacco.

Today, Dr. Leone does employ receptor biology, using nicotine replacement to help people quit smoking. But deep down, with his focus on why people have a dependence on tobacco, he firmly believes that tobacco addiction, and all addictions, have a component of disordered learning. He also believes there is simply too much disordered learning in our society for us to make significant inroads into our multiple addictive behaviors.

There is a lot of evidence to support Dr. Leone's position that there is no end in sight to our addictive behaviors. One of the most obvious signs is the epidemic of electronic cigarette use, especially among teenagers. A 2019 survey published by the CDC showed that 31.2 percent of high school students (4.7 million) and 12.5 percent of middle school students (1.2 million) currently use tobacco, the vast majority using electronic cigarettes.[18] Many of these teens have reported trying to quit, but many others have reported seeing no harm in intermittent use. Prevalence of tobacco use among young adults aged eighteen to twenty-four is also increasing, with the rate of 5.1 percent in 2014 increasing to 7.6 percent in 2018.[19] These numbers show that the recently improved smoking rates among our youth have been wiped out.

Not unexpectedly, a crisis of the lungs has emerged because of this new and completely unregulated industry of e-cigarettes and vaping. Beginning in March 2019, the CDC received reports of a few cases of lung injury requiring hospitalization suspected to have resulted from e-cigarettes or vaping. The number of cases skyrocketed in July 2019, with cases reported from every single state, and hundreds in total. The new disease has been termed EVALI (E-cigarette or Vaping Associated Lung Injury), and 77 percent of the cases reported are in people under thirty-five. As of December 2019, forty-eight deaths have been recorded, likely an underestimate.[20] The additive vitamin E may be to blame, but the only way to assure safety is total avoidance of these products.

Beyond nicotine, we are all aware of the eruption of drug overdose deaths in the United States. This crisis has been fueled by the increased potency of available opiate drugs, most notably fentanyl, but these new drugs are only part of the problem. Overall, there has been an increased use of illicit drugs in the United States, from 8.3 percent of adults in 2002 to 11.2 percent in 2017.[21,22] And with the easy availability of all types of drugs today, both legal and illegal, Frank Leone is not sanguine about any end to addiction in our society. The mass legalization of cannabis for recreational use will certainly not help, as this drug, too, can teach our brain to crave inappropriate rewards, just like any other drug.

The solutions may be obvious, to focus on the personal and societal causes of our addictions, but until those issues are addressed, the opiate crisis, the tobacco crisis, and the electronic cigarette crisis will be with us in one form or another, and these addictions will continue as one of the greatest threats to the health of our lungs.

Chapter 8

Health Is Not the Absence of Disease: Climate Change

The advent of the use of fire in human societies several hundred thousand years ago was a huge breakthrough; fire provided warmth, light, protection, and fuel for cooking. This knowledge led directly to an improved diet with far fewer germs, the development of our brains since nutrition was vastly improved, and the subsequent ability to colonize the planet. The benefit to humanity of creating energy on demand has been immeasurable.[1]

The one downside in regard to our health is the stress energy production has put on our lungs. We've all seen the pictures of what our lungs are up against: billowing fumes wafting out of smokestacks, cities enshrouded in toxic air, people wearing respirator masks as they go about their business. One cyclist in London posted images online of blackened filters from a mask he wore for a few days while riding around the center of town.

The facts back up what we all sense is happening. The World Health Organization (WHO), along with the well-respected Commission on Pollution and Health, have outlined the problem in stark detail. In an average year, air pollution causes 6.5 million premature deaths, more than 90 percent of them in developing countries.[2,3] All forms of pollution contribute to one out of every six deaths worldwide; in the most affected countries, this number increases to one out of four deaths. As has been said, postal code is often more important than genetic code in determining health and life expectancy.

Alarming statistics for poorer countries tend to give those in wealthier countries a false sense of security. The truth is that every nation is affected, with 91 percent of the world's population exposed to air that is of substandard quality, including many urban dwellers in those wealthier nations. In all communities, the weakest and most vulnerable suffer most: mortality from lung diseases is concentrated in those under five years of age and those over sixty. And according to the 2018 report by the Commission on Pollution and Health, published in the journal *Lancet*, these mortality numbers may be vastly underestimated. We simply do not know all of the adverse health effects of air pollution.

Indeed, the problem of noxious air worldwide is getting worse, not better. This grim reality was summed up in chilling fashion in the Commission's *Lancet* article: "Pollution is one of the great existential challenges of the Anthropocene epoch . . . [It] endangers the stability of the Earth's support systems and threatens the continuing survival of human societies."[4]

One of my first brushes with the effects of air pollution came in 2009, when I was working in the pediatric pulmonary division at the Red Cross Children's Hospital in Cape Town, South Africa. Each morning, my colleagues and I would do rounds in the pediatric ICU. One morning, we stopped to see an eighteen-month-old girl who lived in a local township and had been admitted to the hospital overnight for pneumonia. Her name was Lisedi, and she lay on a little bed in an open room, with an oxygen mask strapped to her nose, her big eyes looking up at us with trepidation. Her mother sat nearby, anxious but quiet as the attending doctor explained that Lisedi was receiving some powerful antibiotics to fight her infection.

"That little girl worries me," Max Klein, the attending physician, commented before seeing more patients in another part of the hospital. "Her breathing isn't right for somebody who's already had a few doses of antibiotics."

I nodded my head, accepting at the time that Lisedi's fight was simply between her lungs and the bacteria she had recently picked up.

In reality, her struggle had likely started years ago as we now know that toxic air exposure in children is a significant contributor to respiratory infection acquisition, especially in children like Lisedi, who live in low-income areas. Children are more susceptible to the effects of pollution, indoor and outdoor, for numerous reasons, including the fact that their growing airways are more permeable to particles, and they cannot metabolize and detoxify these particles as well as adults. The end result is that lower respiratory tract infections are the number one cause of mortality in children under five years of age, accounting for some 570,000 deaths each year. While the immediate cause of Lisedi's illness was a bacterial infection, the air she was breathing was what allowed those bacteria to penetrate her lungs.

The morning after seeing Lisedi in the pediatric ICU, Max Klein and I attended the weekly radiology conference at the Red Cross Children's Hospital. The room was dimmed for adequate viewing, and the radiologist started methodically putting up the morning's X-rays for review. All was normal until we stopped at a highly unusual chest X-ray of a very young child. Nobody mentioned the name, but based on the patient's age and history, we knew it was Lisedi. Instead of the normal lung tissue, we saw huge bubble-like structures. Max quickly asked the radiologist to compare the X-ray to the one taken the day prior, which showed no bubbles. There was some debate on what was happening until Max shut it down by saying, "She's clearly got an infection in her pleural space. That's the only thing that could give you those big changes in such a short period of time. That can't be in the lung."

The pleural space is the area between the chest wall and the lungs, normally filled with a small amount of lubricant fluid. In this patient's case, that space was now filled with bacteria that were causing intense inflammation, as well as producing gas, giving the appearance of bubbles, that compressed the lung. Only one intervention could be done, as Max said next: "She needs two chest tubes for drainage. And she needs them right now." A principle of curing microbial disease is that infections need a place to drain. The pleural space is closed, and it needed to be opened up by inserting a tube.

One senior doctor hurried to the ICU. I followed with one of the junior physicians. Upstairs, we saw little Lisedi, struggling mightily with her breathing. Alarm bells were going off intermittently on her monitor, as she was now unable to get enough oxygen into her body. The two doctors inserted the chest tubes, and with each tube insertion a huge rush of gas came out, gas produced by the bacteria.

However, after the second tube was placed, the situation worsened. Although necessary, the tubes had upset whatever delicate balance had existed in Lisedi's body. She succumbed to an irreversible cardiac arrest soon after and passed. Bacterial pneumonia was what would be listed as her cause of death, but given where she was from, that probably wasn't telling the whole story.

The history of air pollution and concomitant lung disease likely stretches back to the first instances of organized human society, some forty thousand years ago. These early effects were limited to the burning of wood for fire, and while this practice had little impact on our environment in terms of climate change, the short-term and long-term problems associated with wood burning were present then and remain issues today.

Burning wood has the potential to release not only noxious gases, such as carbon monoxide (CO), hydrogen cyanide, and ammonia, but also particulate matter (PM). Of all the types of air pollution, PM is thought to be particularly devastating to our respiratory system. It is divided into three sizes, measured in micrometers (one millionth of a meter, abbreviated as μm): coarse matter, which is less than 10 μm (PM10); fine matter, which is less than 2.5 μm (PM2.5); and ultrafine matter, which is less than 0.1 μm. For comparison, fine beach sand is about 90 μm, and the diameter of a human hair is about 70 μm.[5]

All PM is potentially harmful, but fine and ultrafine matter are thought to cause most critical health issues. Coarse PM10 particles, derived mostly from natural sources, such as soil and sea salt, are small but generally still big enough to be handled by the defense systems of the nose and upper part of the lungs and coughed or sneezed out before they reach the deep alveoli. On the other hand, PM2.5 and smaller particles can lodge deep in the lungs and cause detrimental inflammatory reactions.

Wood burning releases PM2.5, most significantly in the form of partially burned carbon. There is evidence that mummified lung tissue from ancient Egypt, Great Britain, and Peru had blackened areas, likely from wood burning. In ancient Rome, PM was such a major problem that writers coined the terms *gravioris caeli* (heavy air) and *infamis aer* (infamous air) to describe the clouds of pollution that enveloped their city. In AD 61, upon the improvement of his health after leaving Rome, Seneca wrote, "No sooner had I left behind the oppressive atmosphere of the city . . . the smoking cookers . . . clouds of ashes . . . [and] poisonous fumes, than I noticed the change in my condition."[6]

Air pollution continued unabated through the centuries. One dramatic example is that of Great Britain, where people began to burn coal in the twelfth century, as the country started to run out of trees. This was sea-coal, which washes up on the beaches of Great Britain from underwater sources that become eroded by tides. It appears most abundantly on the beaches in the north; in the past, it was brought to London and burned in large amounts. With this new source of energy, the smoke from burning coal mixed with fog to blanket the city.

Sea-coal is a particularly noxious form of coal, releasing high levels of sulfurous smoke when burned. Through the centuries, various British kings tried to limit coal burning, but with little success. In 1306, Edward I banned the burning of sea-coal and tried various punishments to enforce the ban. He imposed large fines and destroyed furnaces, and even threatened the death penalty. A few people were tortured—one was executed—but England's citizens continued to burn coal in vast quantities. In 1661, Charles II tried a more subtle approach when he employed the author John Evelyn to write a book about the effects of burning coal and other substances. In his book *Fumifugium*, Evelyn appealed to ancient wisdom about the power of the breath, writing, "the *Philosophers* have named the *Aer* the *Vehicle of the Soul*, as well as of the Earth, and this frail Vessell of ours which contains it; since we all of us finde the benefit which we derive from it." The inhabitants of London, he continued, "breathe nothing but an impure and thick Mist, accompanied with a fuliginous and filthy vapour, which renders them obnoxious to a thousand inconveniences, corrupting the Lungs, and disordering the entire habit of their Bodies." After documenting in part one of his book that "London is hell," in parts two and three he recommended solutions, such as moving the sources of pollution out of the city and establishing gardens with flowers and other vegetation within city limits.[7]

Of course, Evelyn was ignored. With the advent of the Industrial Revolution in the latter half of the eighteenth century, coal was burned in massive amounts to power thousands of factories and locomotives. The factories were often located in urban areas, and their spewing furnaces blackened both the air and the water of industrial cities. J. G. May, an observer from Europe sent to report on England's factories in 1814, vividly described the situation in Manchester: "There are hundreds of factories in Manchester which are five or six stories high. At the side of each factory there is a great chimney which belches forth black smoke and indicates the presence of the powerful steam engines. The smoke from the chimneys forms a great cloud which can be seen for miles around the town. The houses have become black on account of the smoke."[8]

Despite the occasional protest and the obvious decline in air quality, the burning of coal continued unchecked for centuries throughout Europe and, later, in the United States. In Great Britain, the matter reached a tipping point in December 1952 with the Great Smog of London. For five days, a thick toxic cloud hung in the air, and the city endured a pollution like none of the previous "pea-soupers." Part of the problem was the cold weather, which caused people to burn more coal. Coal-fired power stations, vehicle exhaust fumes, diesel buses, and steam locomotives also contributed to the pollution.

But what really tipped the balance toward a medical tragedy was a weather-related event called a temperature inversion. Normally, the warmest air is closest to the Earth, as solar radiation is absorbed and heats the air just above us. Warm air naturally rises, causing cold air above to rush in, creating wind and diluting out pollution. However, when certain meteorological conditions are met—in the 1952 London case it was the lack of wind—a warm layer of air rises above a cold layer, the cold air does not move or rise, and all of the pollution and smog becomes trapped.

So, during those few days in London, the trapped pollution just sat on top of the city, creating chaos. Visibility was minimal, making driving impossible and shutting down not only personal vehicles but buses and taxis as well. Some people died, not because of lung illness, but because they couldn't see and fell into the Thames River and drowned. Smog (smoke combined with fog) seeped into buildings. Sadler's Wells Theatre had to close after the first act of Puccini's *La Traviata* due to poor visibility and air quality. Wembley postponed a soccer match, and there were press reports of cows asphyxiating in the fields. Worst of all, four thousand people died in the immediate aftermath, and eight thousand additional deaths during the following January and February were likely related. Undertakers ran out of caskets, and florists had no more flowers. Many thousands more suffered from illnesses because of the devastating event, and the people of Britain were in shock. Their air had been taken away from them in a dramatic fashion.[9]

The United States had a similar air pollution disaster a few years prior to the London incident. Donora, Pennsylvania, lies in a valley along the Monongahela River, about thirty miles south of Pittsburgh. In the 1940s, it was the site of two major industrial plants, Donora Zinc Works and American Steel & Wire, both owned by United States Steel. From October 27 to October 31, 1948, an inversion similar to the one that would later occur in London gripped the town, trapping dense smog and pollution.

The citizens of Donora began falling ill en masse, as deadly levels of pollutants, including sulfuric acid, nitrogen dioxide, and fluorine, built up in their lungs, choking them with lethal gases and subsequent inflammation. Doctors' phones rang off the hook until an emergency call center was set up in the town hall. The nurse at the steel mill, Eileen Loftus, painted a tragic picture of the first afflicted workers she treated:

"He was gasping. I had him lie down and gave him oxygen. Then another man came in, and another."[10] It was difficult to reach the patients as visibility shrank, making driving almost impossible. The Halloween parade was a truly ghostly affair, and the football team abandoned its passing game because of poor visibility. All told, twenty people died in the immediate aftermath, and fifty more succumbed the following month. Nearly half of the fourteen thousand residents fell ill. One of the fatalities was baseball Hall of Famer Stan Musial's father, fifty-eight-year-old Lukasz Musial. Over the following decade, the mortality rate in Donora remained inordinately high.

Both of these catastrophes represented turning points in a drive toward cleaner air. In Donora, lawsuits were filed, and awareness was raised. The tragedy caught the attention of President Harry S. Truman, who in 1950 convened the United States Technical Conference on Air Pollution and referred to Donora in his conference invocation. This event paved the way for the 1955 Air Pollution Control Act and later the 1970 Clean Air Act, which set up strict regulations at state and federal levels to limit emissions from industrial and mobile (cars and trucks) sources. Great Britain followed a similar path with its own Clear Air Act of 1956 and a transition from coal as the nation's primary source of energy.

The legally mandated changes have undoubtedly saved many thousands of lives, but today we are still struggling with poor air quality on an unprecedented scale, both in the United States and worldwide. Since the days of coal and wood burning, we have become dependent on other fuel sources, such as oil and gasoline, which release their own toxic pollutants. The United States Environmental Protection Agency has identified six types of pollutants with serious effects on human health—PM, ozone (O_3), sulfur oxides (SO_2), nitrogen oxides (NOx), carbon monoxide (CO), and lead.

The major producers of these six pollutants are power plants and car engines. Other important sources, such as agriculture, are not as frequently discussed. A 2015 study published in *Nature* noted that, in an average year, the United States had 16,929 deaths attributable to polluted air from power plants, while 16,221 deaths were attributable to pollution from agriculture.[11] Modern farming is a huge producer of toxic PM, largely from ammonia from fertilizer use and animal waste. This ammonia combines with nitrogen from car exhaust fumes and sulfate from power plants to form deadly PM2.5. This is what regularly puts the Fresno–Madera city area in California among the top five most-polluted cities in the country. Their size and long growing season put farms in California high on the list of polluters, but in the summertime, farms of the American Midwest produce up to 40 percent of the total emissions measured in their states.[12]

The smell of wood-burning stoves reminds us of crisp fall days, and the practice is generally regarded as a harmless way to heat our homes. But these stoves release massive amounts of PM, benzene, and formaldehyde into the air, which can travel for miles. The inhalant is no better than, or even different from, cigarette smoke, largely because most wood stoves burn fuel incompletely and inefficiently. The use of wood-burning stoves is so prevalent in some states that it is often a major contributor to PM pollution. It has been estimated that every winter in Washington State, residential wood stoves contribute 35 percent of the total small-particle pollution—the single largest contributor—10 percent more than agricultural dust and almost twice as much as the fumes emitted by cars and trucks.[13] The situation in Great Britain is similar, with the PM2.5 from wood-burning stoves contributing more than twice the amount of particle pollution than automobile traffic. (The toxicity of automobiles lies more in their output of the gases nitrogen oxide, carbon monoxide, carbon dioxide, and sulfur dioxide.)[14]

With all of the things that we burn and consume for energy purposes today, according to the 2019 American Lung Association's *State of the Air* report, a staggering 141 million Americans are exposed to unhealthy levels of air pollution, about 43 percent of the population.[15] This is an increase over the numbers reported in the prior two years, a warning that we are headed in the wrong direction after decades of progress. American cities in the West, especially in California, known for outdoor activities and a healthy lifestyle, dominate the lists of most polluted cities. In the United States, Los Angeles ranks number one in ozone pollution, Bakersfield is number one in short-term particle pollution, and Fresno–Madera–Hanford is number one for year-round particle pollution. Geography is part of the problem in the West, with mountains often blocking gases that would normally disperse. Places such as Salt Lake City, surrounded by the Rockies, frequently experience temperature inversions during the winter, causing unhealthy levels of PM, ozone, and nitrogen dioxide. This phenomenon triggers a "Mandatory Action Day," when wood and coal stoves are not to be used, along with fire pits, fire rings, and campfires. Residents are also asked to carpool, use public transportation, and consolidate trips when possible.

The worldwide data is even more concerning. As mentioned, the WHO estimates that 91 percent of the global population lives in places where air quality guidelines are not being met. Regulations, such as those in the 1970 Clean Air Act, are nonexistent in parts of Eastern Europe, in Russia, and throughout the developing world. Simple protective devices such as filters and scrubbers on top of smokestacks are not used, and basic laws governing exhaust from cars and trucks are absent. For this reason, people are exposed to a complex mix of solid and gaseous toxins from vehicle exhaust fumes, road dust, and smokestacks, and this exposure significantly contributes not just to lung diseases such as pneumonia, asthma, and cancer, but also to strokes and heart disease. We are further beginning to understand the effects of pollution on body systems not previously known to be affected: a 2017 study from Columbia University showed a significant link between air quality and the risk of osteoporosis, and a 2019 study from the University of Southern California pointed to a connection between particulate-matter exposure and Alzheimer's disease.[16,17]

Today, in cities such as Beijing and Delhi, levels of PM2.5 regularly reach levels of 300 mcg/m^3 or higher, with a level of 30 being the high end of the safe range. Events reminiscent of The Great Smog of London and the 1948 tragedy of Donora, Pennsylvania, occur on a yearly basis.

In November 2017, levels of PM2.5 exceeded 900 mcg/m³ in Delhi, and four thousand schools were forced to close for almost a week. The chief minister of Delhi, Arvind Kejriwal, called the city a "gas chamber." One chest surgeon commented in the *New York Times*, "I don't see pink lungs even among healthy nonsmoking young people." United Airlines suspended flights into the city, and construction projects were halted.[18] This is sadly a not uncommon event in Delhi, and schools had to close there again in November 2019 due to toxic air.[19] Long-term lethal health effects are sure to follow.

Unsurprisingly, of the 6.5 million deaths caused each year by air pollution (11.6 percent of all global deaths), 55 percent come from China and India, countries with large, dense populations and rapidly expanding economies.[20] The problem is not just outdoor air pollution, but indoor air pollution as well, as households in developing countries regularly burn fuel for cooking and heating, producing toxic gases and PM that linger indoors because of poor ventilation. The source is usually burning wood, dung, coal, or crop waste, referred to as biomass fuels. About three billion people, concentrated in Africa, India, and China, rely on the burning of biomass fuels. In these regions, an estimated four million people die annually from indoor air pollution, with many of these deaths occurring among children under the age of five.[21]

Humans are not the only victims of all of this noxious air; it also pollutes oceans, poisons trees, and of course contributes to global climate change, which in turn makes fixing these issues even harder as the two problems feed off each other.

As physicians, we are tasked with addressing the downstream effects of dangerous trends in society, such as the use of tobacco or the worsening quality of our air. One physician pushing back against this status quo is Heather Zar, a doctor on the faculty at the Red Cross Children's Hospital in Cape Town, who through innovative research is bridging not only the wide gap between public policy and the role of doctors, but also the gaps between rich and poor, and between developed and developing countries.[22]

The Drakenstein Child Health Study, which began in 2012, is Dr. Zar's latest and most ambitious undertaking, and aims to shine a bright spotlight on what pollution is doing not only to our lungs, but to our brains, to our immune systems, and even to the bacteria that colonize us.[23] A main goal of the study is to understand why pneumonia is the leading cause, worldwide, of both death and illness among children under five years old, and to figure out what can be done about it. If Dr. Zar is successful, we'll learn how to prevent bouts of childhood pneumonia like the one that killed little Lisedi.

Drakenstein is an area just inland from Cape Town, on South Africa's eastern coast, and as in most of the country, many of the inhabitants of the region are poor, live in a semi-urban environment, and are exposed to a multitude of indoor pollutants and infectious agents. Dr. Zar and her colleagues chose Drakenstein, but in reality, the study could have been conducted in almost any part of South Africa, or in any other African nation: the continent accounts for just 18 percent of the global under-five population but 42 percent of total deaths for this age group each year.

Dr. Zar and her colleagues decided to study the children even before they were born. Potential effects on a child's lungs are known to start in utero, so Dr. Zar's team enrolled mothers who were twenty to twenty-eight weeks into their pregnancies, with a plan to study their children's habits up to the age of five, as well as the habits of their mothers and their entire households.

The usual suspects for harming children's health are present in Drakenstein's homes: the buildup of indoor pollution from smoking tobacco and cooking with biomass fuel. Dr. Zar's team is also studying the nutrition of the mothers and their babies, the genetics of the babies and their parents, and psychosocial issues that different families deal with. Finally, the team is analyzing the microbiome of the children, a new area of inquiry that is emerging in the study of childhood pneumonia.

A term that emerged in the late 1990s, *microbiome* is defined as the collection of microorganisms that live in a particular environment, including in and on humans. We have always known about bacteria living in the human gut and on the skin, but newer molecular techniques have allowed us to catalog the vast number and scale of organisms living in every organ of the human body. Overall, some ten thousand trillion organisms live in every human; for each one of our cells, there is a microbial cell which lives in and on us.[24] The vast majority of these organisms live in the large intestine, but some inhabit organs previously thought to be sterile, such as the bladder and the lungs.

Today, we know that hundreds of species of bacteria colonize our lungs, including *Provatella*, *Fusobacterium*, and *Streptococcus*, along with fungi, such as *Candida* and *Saccharomyces*. Many of these bacteria and fungi clearly perform important functions, primarily keeping other harmful bacteria out by producing inflammatory proteins that both kill invading bacteria and induce the lung cells to produce bacteria-fighting proteins.[25]

Knowledge of the microbiome is forcing scientists to rethink how lung diseases occur as well as how they might be treated. Those patients with COPD, cystic fibrosis, and asthma have all been shown to have very different lung bacteria compared with subjects without lung disease, likely making them more susceptible to other seasonal infections. We also know that exposure to household air pollution significantly changes the populations of bacteria in the human lungs, and smoking alters this microbiome in the lungs, nose, and throat. Dr. Zar and her colleagues want to find out whether disruption of the lung microbiome from exposure to pollutants is the primary mechanism by which harmful bacteria are able to cause infections in children with pneumonia. To this end they are culturing the bacteria in the lungs and noses of their young subjects and matching the results to levels of pollutants in their environment.

With 1,140 mother-child pairs having completed one full year in the study in 2016, some definitive results of the Drakenstein Child Health Study have already been reported. They showed that many children were being raised in a toxic environment—one-third of the women smoked while pregnant, and 56 percent of the newborns had detectable cotine (a nicotine byproduct) levels in their urine samples. There were also high levels of biomass fuel exposure in the homes. The rates of pneumonia were elevated in the cohort as a whole, despite all the babies being appropriately immunized. Among the babies who contracted lung infections, their lung function, measured at one year of age, was lower than that of children who were able to remain infection free (adults with pneumonia generally recover all of their previous lung function, barring a very severe infection).[26]

Decreased lung function is known to put children at increased risk of contracting pneumonia again, as well as developing asthma. But there are other implications for these children beyond the lungs. Data clearly shows that decreased pulmonary function in adults leads to more dementia and cognitive impairment later in life. Some findings on this effect in children are emerging, too, and the Drakenstein Study team plans to add to this information by examining MRI brain scans of some of their subjects to find out if respiratory infections and pollution influence brain development. If the answer is yes, it would provide another important reason to start cleaning up the air that we all breathe.

The dangers we face from indoor and outdoor air pollution are staggering, given the huge number of people exposed to noxious air. The good news is that there are steps we can take right now, and we know that by taking these steps we can improve the situation immensely. This includes creating cleaner cars, building cleaner power plants, and burning carbon-based fuels, such as coal and wood, to completion to minimize smoldering output.

We have begun many of these initiatives in the United States, and they have been working. Despite setbacks in air quality in the past few years, since 1970 emissions of the six most common pollutants have dropped by 70 percent, even while our population, economy, and energy use have all significantly increased. Equally remarkable is the documented effect of cleaning up the air on human lung function and development. In a study published in 2015 in the *New England Journal of Medicine*, three cohorts of children in Los Angeles had their lung function checked annually over a four-year period, starting at age eleven.

The first group started in 1994, followed by a second group in 1997, and a third group in 2007. Throughout this period, the air quality in Los Angeles improved significantly. The results show that for the later cohorts of kids, average increase in pulmonary capacity over the four-year period was greater. With fewer noxious chemicals, their lungs were able to grow larger, which will undoubtedly lead to longer life spans in the future.[27]

Other studies have demonstrated this very tight correlation between local pollution levels and health outcomes. The Utah Valley is an area of Central Utah where low smoking rates are the norm. The Geneva Steel mill, however, was a significant source of PM10 in the area for several decades from when it opened in 1944 until its closure in 2001. A strike occurred during the winter of 1986–87, and researchers used this opportunity to measure the effects of PM10 on local health outcomes. Compared to other winters, levels of PM10 during the winter of 1986–87 were much lower in the Utah Valley. Also significantly lower during the period of the strike were hospital admissions of children for asthma, bronchitis, and pneumonia: there were seventy-eight admissions for asthma and bronchitis during both the winter of 1985–86 and the winter of 1987–88, while during the strike year the total was only twenty-three.[28] A similar study in the Southwest analyzed health effects of a copper smelters' strike that lasted from July 1967 to April 1968 and affected New Mexico, Arizona, Utah, and Nevada. The mortality rate in the region dropped 2.5 percent because of the strike.[29]

Isaac Newton's third law of motion states that for every action, there is an equal and opposite reaction. For a long time, breathing polluted air led to declining lung health. We now have proof that the equation can be reversed—an intelligent attempt to clean up our air can have widely positive effects on the lungs and overall health.

Nonetheless, experts have warned that controlling pollution effectively over the long term lies in how we generate energy to heat our homes, drive our cars, and produce goods and services. We will need to embrace renewable sources of clean energy, such as solar, wind, and hydropower, along with clean use of biomass and geothermal sources. Arguments against this approach, that it will cost too much money, or that it will impede progress, have been shown to be false. The United States and other countries have received extraordinarily positive economic returns from a concerted effort to clean up polluted air.

The world has focused most recently on climate change and reduction of carbon emissions. The Kyoto Protocol began international cooperation on this issue in 1997, then was replaced by the Paris Agreement in 2015. As of 2019, 194 nations have signed on, with the United States initially on board. This changed in November of 2019, when the United States federal government signaled its plan to withdraw after a one year waiting period, throwing the potential success of global cooperation on this looming environmental disaster in doubt.

Air pollution specifically has traditionally been dealt with at the individual country level. In the United States the Environmental Protection Agency has handled the majority of issues, while in Europe the European Environment Agency has set standards. Both regions have seen improvements in air quality over the decades, but as mentioned, many areas are still exposed to substandard air quality. Air pollution can be difficult to regulate with one agency or agreement, because so many different sources contribute: agriculture, road transport, energy production, natural phenomena, local businesses, and households. Successful international cooperation on an environmental issue is not without precedent, however, as demonstrated by the hugely successful Montreal Protocol, which in 1987 laid out a plan to limit ozone-depleting chlorofluorocarbons (CFCs).[30] With the entire world on board, a healing ozone layer has been seen every year since.

Outside of the United States and Europe, other countries have taken on the issue of air quality with variable success. China in 2013 issued the Air Pollution Prevention and Control Action Plan, which successfully reduced PM2.5, PM10, and toxic gasses in seventy-four cities, with an estimated 47,240 lives saved between 2013 and 2017.[31] Significant problems with air quality still exist in China, but this initiative is encouraging. India passed the Air (Prevention and Control of Pollution) Act in 1981, but despite this legislation, air quality has steadily worsened over the decades, and India is home to twenty-two of the world's thirty most polluted cities, with an estimated 141 million people breathing air that is ten times more noxious than WHO limits.[32,33]

In the United States, leadership at the federal level has recently been lacking, but many states are showing the way forward. With respect to clean energy generation, results are impressive. Kansas, Iowa, and Oklahoma are leading the way in wind energy production, with, respectively, 36, 34, and 32 percent of their total utility power generation from wind in 2018.[34] Texas, number one in the country in total wattage produced, now has more wind capacity than coal-fired capacity in the state. Renewable energy providers in California produced 34 percent of the electricity used in 2018, with solar power accounting for 10 percent of that total.[35] Mississippi added enough solar energy in 2017 to power twenty-five thousand homes every year.[36]

Existing clean energy sources will likely not be enough to replace fossil fuels completely, and new technologies will need to be developed. This is happening today with the production of biogas from animal-waste breakdown and the capturing of energy from the ocean. Advances are coming in fusion energy technology, which involves the combination of two lighter atomic nuclei into one, with the subsequent release of energy (in contrast to nuclear fission energy).

Despite these promising innovations, air pollution remains a problem, and polluters need to be continually monitored. The warning offered in the American Lung Association's 2019 *State of the Air* report should not be ignored. It postulated that air quality in the United States has deteriorated in the past few years not because of specific pollutants from man-made sources, but from the growing plague of extreme wildfires. In the United States, wildfires have been concentrated in California, where 2017 and 2018 saw the deadliest and most destructive fires in the history of the state, destroying thousands of houses and causing many to flee.[37] The fires in the Amazon rainforest and then in Australia horrified people throughout the world in 2019 and 2020, and the health effects of these two events will be felt for years. Given that climate change is fueling these fires, the frequency and lethality of these events are expected to increase.

Addressing climate change and embracing clean energy sources will be a major part of our effort to improve air quality after recent setbacks. In the face of such enormous challenges, the goal of clean energy sounds almost quaint. But if we manage to make a total commitment to clean energy sources, we could accomplish something for the first time in the thousands of years of our civilized existence: living an advanced lifestyle without polluting the air that we breathe. And if we are able to achieve this goal, we will go a long way toward keeping our lungs and bodies healthy.

Chapter 9

Exposures Unnecessary: Time Does Not Heal All Wounds

M uch is mysterious about the functioning of the lungs, and nothing more so than the reactions initiated after an inhalation. The lungs have a dual purpose, and the purposes unfortunately are in direct opposition to each other—bringing oxygen in while keeping everything else out. The latter objective is almost impossible, especially since we take more than fifteen thousand breaths a day.

The lungs have an advanced system for both keeping noxious particles out and rejecting unwanted particles if they do get in. The system of defense begins in our nose, where hairs help filter the air. The airways, including the bronchi and bronchioles, also have tiny hairs, termed *cilia*, which beat constantly to expel unwanted particles that make it past the initial defenses. The other lines of protection we have are coughing, sneezing, and clearing our throats, which all serve to manually force noxious material out.

The defenses of our lungs are not foolproof, and dust regularly gets down into our airways. Proportion matters here, and the smaller the particles are, the farther down they can penetrate, with those measuring five microns or less able to get to our alveoli. In an average room, every cubic foot may contain twenty thousand to thirty thousand particles of respirable dust in this size range. Construction sites may contain eight hundred thousand particles of respirable dust per cubic foot. It is impossible to keep all of these particles out.

Typical home dust is made up of a mixture of tracked-in soil, soot, ash, particulate matter from cooking, dust mite debris, flakes of human skin, and lint from clothing and bedding. The makeup of the outside air is determined by where you live, but is usually a mix of soil particles, pollen, particles from exhaust, and perhaps salt flecks if you live near the ocean, or sand particles if you live near a desert. The composition of air at workplaces very much depends on the type of worksite and its location.

We have evolved to live with dust, and most of it is harmless. Dust also does a lot of good in nature, like absorbing water vapor, which otherwise would make the biosphere extremely muggy. Fungal spores travel as dust, landing to perform the all-important function of breaking down dead matter. Pollen has the critical task of fertilizing flowers and plants, a job done with the help of bees but also by pollen in the form of free-floating dust.[1] Farm dust is thought to help protect children from developing asthma and allergies later in life, a salutary interaction between our immune system and dust.

In addition to all this beneficial dust, however, there is also a lot of bad dust, especially at our worksites, where we often have little control over it. And since our worksites have changed so radically in the past two centuries—even in the past twenty years—keeping up with potential airborne threats is hard. Of all the organs in the body, the lungs, on the forefront of our interaction with the environment, are the most affected by toxins in the workplace.

Police officer Cesar Borja was one of the forty thousand American heroes who selflessly went to Ground Zero in New York City to aid in the cleanup effort after 9/11. He worked there sixteen hours a day, coming home sweaty and grimy and tired. He stayed at home only long enough to eat, shower, and sleep a few hours, and then returned to help with the massive effort.

The damage was vast, with four hundred million pounds of steel and six hundred thousand square feet of broken glass spread over sixteen acres in a toxic pile of burning rubble. On the job, masks were optional. Some workers wore them, but many—including Cesar—did not. Little

did he and many other workers know that the air they were breathing was poisoned with toxic dust particles such as asbestos, mercury,
lead, and cadmium. The fires that burned on the site for months added
dioxins and polycyclic aromatic hydrocarbons, both known carcinogens.
Particles of gold from melted jewelry were also in the air.[2]

Cesar worked at Ground Zero for three months. Five years later he
died in a hospital bed from idiopathic pulmonary fibrosis, the devastating illness that causes scarring in the lungs. His family then asked the
questions that came to everybody's mind: Did his working at Ground
Zero without a mask cause his illness? If Cesar Borja didn't know he
was inhaling a toxic mixture of carcinogens, were there people who did?

The use of respiratory masks at the 9/11 site is a complicated
matter, not surprisingly, given the uniqueness of the massive cleanup
effort. The first firefighters who arrived at the cleanup site used their
own full-face masks and breathing apparatus, but the cartridges lasted
only a day. Firefighter Palmer Doyle from Brooklyn was at the site on
the first day, and then arrived back at the scene on September 15 with
fifty other firefighters. He remembers being handed one respirator
for the group. It was given to the youngest member, and they started
their work. An order was placed by the fire department to the city on
September 28 for five thousand masks, which cost about fifty dollars
each, and ten thousand replacement cartridges. The order went unfilled
for two months.[3]

Some 150,000 masks were distributed during the cleanup effort,
but their use was sporadic, and they often didn't fit well. Those working for the Pentagon at the Ground Zero site were escorted off if they
weren't wearing a mask, while New York's city and state workers were
allowed to work without them. Some wore paper masks or surgical
masks, but these were later deemed to be useless. The effective P100
masks, outfitted with cartridges that block almost all particulate matter, were uncomfortably hot and made communication among workers
difficult. Many reasoned they could do without masks.

This sentiment was reinforced by Christine Todd Whitman, head
of the EPA at the time. Three days after the attacks, she stated, "The

good news continues to be that air samples we have taken have all been at levels that cause us no concern."[4] She did state that those working directly on the pile should take precautions, but raised no further alarm. This line was echoed many times by the mayor of New York City, Rudolph Giuliani.[5]

The issue at the time was complicated. The World Trade Center area was not just a cleanup site, it was at first a massive rescue site, a fire that needed controlling, and also a crime scene. Many firefighters and EMT workers naturally felt that by donning a mask they were selfishly putting their own health above those who needed rescuing. Workers also wanted to show the world that America would not be deterred.

But the story of the World Trade Center cleanup now reads like an uncontrolled science experiment. Tens of thousands of people were exposed to all manner of toxic dust, costing billions of dollars in lawsuits and future medical treatment, to say nothing of the diseases incurred and the suffering they caused. Environmental health expert Dr. Paul J. Lioy, a scientist at New Jersey's Environmental and Occupational Health Sciences Institute, watched the buildings come down on his television at home and immediately knew it could be a health disaster. Similarly, occupational physicians at Mount Sinai School of Medicine, among the most knowledgeable experts in the world on the subject of dust-induced illness, instantly recognized the tremendous potential for health issues.

At Fresh Kills, the Staten Island dump where 9/11 waste was deposited, there was a strict mask policy, with a more than 90 percent compliance rate. At Ground Zero, by contrast, on any given day as few as 30 percent of workers wore masks. Later, with plenty of toxic dust still inside local residences and businesses, on bookcases and coffee tables, and under beds and desks, families came back to their homes, and workers went back to their offices. The different types and manifestations of lung disease we are seeing today because of this exposure are tragic.

Cesar Borja finished his work on the 9/11 cleanup site and went back to his job at the New York Police Department tow pound. Initially, things seemed fine, and he retired in 2003, But the quiet didn't last long. In early 2005, he started noticing a cough, dry and incessant. He went to see a doctor, and a chest X-ray and CT scan followed. He was diagnosed with pulmonary fibrosis of the lungs, an irreversible condition in which fibrous tissue is produced in the lungs at an increasing rate, turning them from a soft sponge to dense rock, usually over the course of months to a few years.

Cesar held his own for a few months, until he woke up one morning unable to breathe. His wife called an ambulance, and at Mount Sinai Hospital in New York City, Cesar was put on a ventilator and given medicines, without positive results. During this time, Cesar Borja Jr. wrote to the local papers about his father's condition, hoping to start a groundswell push for a lung transplant. Normally, patients on ventilators cannot get lung transplants, but Cesar Jr. argued this was a special case.

Cesar's powerful story landed on the front page of the *New York Daily News*.[6] As Cesar languished on a ventilator at Mount Sinai, his family's phone rang off the hook with calls from local and national newspapers and television outlets. Hillary Clinton, then a senator from New York, took notice, and Cesar Jr. found himself at Ground Zero speaking alongside her and others regarding the health problems afflicting first responders. He was then flown to Washington, DC, to be Clinton's guest at the State of the Union address, and later met with President George W. Bush to promote the cause of his father and other injured workers. Sadly, no lung transplant materialized, and Cesar Sr. passed away two hours before the State of the Union address his son attended.

Cesar Borja was one of many afflicted with disease caused by the cleanup efforts after the 9/11 attacks. Dr. Robin Herbert of Mount Sinai Hospital describes three distinct waves that frequently occur after a massive dust exposure. The first is characterized by the acute effects of the inhalant, a burn-like effect. The second wave occurs in the following months, characterized by ongoing inflammation, with diseases such as asthma, or lung conditions that lead to scarring. The third wave, which can happen years and even decades after the exposure, involves cancers and other life-threatening conditions.[13]

The first wave of patients began showing up soon after 9/11, their airways burned by the very caustic Drano-like dust, with a severe dry cough, eye irritation, and cherry-red nasal passages from the burn. The cough from the burn was so harsh, so distinctive, that Dr. Herbert could walk into the waiting room at her clinic and instantly know which patients were from 9/11. Palmer Doyle, the firefighter from Brooklyn who had started work at Ground Zero in the early days, developed a serious case of bronchitis, so severe that his parents didn't recognize his voice when he talked to them on the phone.[8]

Many of these coughs did not resolve easily. Firefighters who showed up on the first day the building collapsed had the highest incidence of this cough syndrome, at 8 percent (128 of 1636), while those who got there on day two had a somewhat lower incidence of 3 percent (187 of 6958). Of those with this cough, 63 percent had evidence of lung function abnormalities.[9] Symptoms of shortness of breath, acid reflux, wheezing, and sinus issues were also higher in these patients.

The second wave of illness began to rise in the following years. Prior to 9/11, the incidence of asthma among those who would work at the World Trade Center (WTC) site was 2.9 percent. By 2002, it was 12.8 percent, and by 2007 it was 19.4 percent.[10] A study published in 2011 looked at nine years of data and showed an asthma rate of 27.6 percent, as well as strikingly elevated rates of sinus inflammation and acid reflux disease.[11] Over the years, unusual diseases such as sarcoid and eosinophilic pneumonia, both inflammatory lung diseases that can cause scarring and respiratory failure, became unusually prevalent. A 2015 review of the data stated that, given the consistency of findings in multiple studies, a causal link between WTC dust and diseases such as asthma, COPD, acid reflux, and scarring diseases of the lung could be made with scientific certainty.[12] High rates of depression and panic disorders among first responders have also been documented.

In 2007, Dr. Herbert gave an interview to the *New England Journal of Medicine*, outlining the diseases she was seeing, but also sounding an alarm about the third wave of even more deadly diseases to follow, namely cancers.[13] Unfortunately for those affected, more than a decade after this interview, higher-than-normal rates of skin cancers, thyroid cancer, and certain blood cancers such as multiple myeloma are indeed occurring. A chilling review published in the *Journal of the American Medical Association Oncology* in October 2018 predicted higher-than-normal levels of prostate, thyroid, and melanoma cases among WTC workers going forward to 2031, with an overall elevated incidence of the total cancer rate.[14] The third wave of illnesses also includes persistent PTSD, as well as greater risk of heart disease and stroke. The dust, it appears, didn't just go into the lungs, but also triggered inflammation throughout the body. Overall, as of September 2019, the CDC estimated that 15,543 cancer cases are related to 9/11 exposure.[15]

The names are etched on three granite walls, honored as 9/11 heroes. However, these walls are not in the shadow of the Freedom Tower in downtown New York, but rather in a park fifty miles west, in Nesconset, Long Island. In the park, known as the "9/11 Responders Remembered Park," the three walls, named Courage, Honor, and Sacrifice, stand serenely, paying homage to those who died not on 9/11, but after that event, and because of it.

The park's caretaker and planner, John Feal, got the land donated, raised money, and added $130,000 of his own to get the fundraising over the line. He spends ten hours a day, without any salary, vetting people to determine whether their names should go on the wall and whether they should receive benefits from his foundation. To make the difficult decisions about who gets on the wall and who gets benefits, Feal reads obituaries, and talks to families, and sometimes asks tough questions. He always stresses that his goal is to be inclusive and to make sure the people who continue to die because of that day are remembered and honored.

John was at the World Trade Center on the days after the towers came down, working long hours, until September 17, when an eight-thousand-pound steel beam fell and mangled his left foot, causing injuries to his spine and knees as well. He lives with chronic pain today. Ironically, he considers himself lucky, because he knows that if he had worked at the site longer, he might be one of the names on the wall he helped build.

John's FealGood Foundation advocates for all the needs of those suffering with 9/11 illness—medications and prescriptions, transportation to chemotherapy sessions and other medical treatments, help with nutrition support and utility costs. The foundation also puts people in touch with specialized doctors and lawyers, and considers other requests on a case-by-case basis. Feal is both honoring those who have fallen and trying to prevent more from ending up on the wall. He has inspired legislation at the state and federal level, making sure the nation lives up to its duty to take care of the countless people who rushed to help in a time of great need.

The topic of occupational diseases rarely attracts attention either in the medical field or the press, except after major disasters or lawsuits. Most of us take for granted that the places where we work are safe, healthy environments with clean air and water. Diseases are something we pick up from viruses, or our genetics, or our lifestyles. Illness does not come from our workplaces.

As a group, occupational lung diseases are less recognized than other types of occupational hazards. The inhalation of toxic dusts can be subtle, frequently going unnoticed because the particles don't always cause a cough or other immediate respiratory symptoms. A worker who was exposed for only a few months might come to his doctor thirty years later with a cancer that has spread throughout the body. Establishing cause and effect can be very difficult.

Once toxic dusts are inhaled, the inflammatory reaction to the particulate matter can also be unpredictable, with some heavily exposed workers developing no disease, while others with limited exposure develop a devastating illness. The same inhaled particle may also cause completely different diseases in different people. Some patients may react by developing asthma, while others end up with pulmonary fibrosis, lung cancer, or other inflammatory conditions. Why the same particle causes different diseases in different people is largely unknown, although it likely goes back to how much is inhaled, how our lungs handle the particle, and our unique genetics.

Occupational lung diseases have been with us for millennia. The lungs of fifteen ancient Egyptian mummies were recently examined and, surprisingly, they show a degree of particulate matter not so different from what is found in the lungs of somebody living in a modern-day city. Ancient Egypt did have dangerous industries, such as metalworking and mining, and citizens were regularly exposed to dust storms. One mummy examined in the early 1970s, a thirty-eight-hundred-year-old named Nekht-ankh, lived to almost sixty, and his lungs contained a remarkable amount of particulate matter, along with fibrosis.[16]

With the Renaissance and modernization in Europe came new occupational lung diseases. As early as 1473, German doctor Ulrich Ellenbog recognized the emergence of these diseases in a book entitled *On the Poisonous Wicked Fumes and Smokes of Metals*. In 1700, the father of occupational medicine, Bernardino Ramazzini, published *De Morbis Artificum Diatriba* (*A Treatise on the Diseases of Workers*). Ramazzini described some two hundred occupational diseases, many of them affecting the lung. Today, the list of workers at risk for occupational lung diseases is surprisingly long. From those who work with seemingly innocuous plants like strawberries, to bakers who work with flour, to the more obvious cement workers, rubber workers, chrome platers, coal miners, and firefighters, workers face real risks from inhalational disease.

Ventilation efforts and personal protective equipment have reduced the incidence of these occupational inhalation diseases. The one exception to this trend, and the inhalational disease with the strangest, most numerous, and most deadly manifestations, is asbestos. There has not been a more deadly inhalation disease in American history than the one caused by this tiny fibrous mineral.

Asbestos is a beautiful fiber, found in soil and rock, that occurs naturally as a long, thin crystal. It is extremely strong, fire resistant, and is found in nature in stunning colors of bright white, emerald green, iridescent blue, and light brown. Although typically thought of as a modern material, asbestos has a long history of use in civilizations going back thousands of years. Twenty-five hundred years ago, the ancient Egyptians used asbestos-reinforced cloth to wrap the bodies of pharaohs for preservation.

Asbestos was also used in ancient Greece and Rome, where it was mined from local quarries. Asbestos was regarded as a miracle fiber and recognized for its flame-resistant properties. The Greek historian Herodotus, in 456 BCE, wrote about bodies wrapped in asbestos shrouds before burning to prevent the ashes mixing with the fire. Romans also wove the fiber into napkins, reportedly throwing them into fires for cleaning, then retrieving them, white and unharmed.[17]

If the Greeks and Romans recognized the unique resiliency of asbestos, they also recognized its potential for harm. The Greek historian Strabo, writing in the first century CE, described a "sickness of the lungs" in those slaves who manufactured asbestos-containing cloths. Pliny the Elder, writing around the same time, also described lung disease in slaves who worked in asbestos mines, noting that they wore masks crafted from the bladder of goats or lambs.[18]

The mining of asbestos declined in Europe over the next thousand years, although its use did not cease entirely. King Charlemagne, as a parlor trick for his guests, would throw an asbestos cloth into a fire and demonstrate his powers by removing it unscathed. Stories abound of traveling hucksters selling asbestos crosses said to be made from the original cross of Jesus, their flame-resistant properties ostensibly proving their divine origin. With the rise of the industrial revolution in the nineteenth century, however, widespread asbestos mining and use began.

Asbestos fiber made its way into a surprising number of products, from building-insulation materials to brake liners of cars to fire blankets, caulking material, vinyl floor tiles, thermal pipe insulation, and cigarette filters; it was even in the snow in the film *The Wizard of Oz*. The use of asbestos continued unabated through the 1970s, despite a loud warning about its danger some fifty years earlier.

Nellie Kershaw worked in the textile industry as a rover at Turner Brothers Asbestos Company in Rochdale, England, taking the raw material of asbestos and spinning it into yarn. She started working there in

1917, at the age of twenty-seven, and began exhibiting respiratory symptoms at twenty-nine. She worked another two years at the mill, until she was unable to carry on. At home, and struggling with her breathing, she wrote a letter to Turner Brothers, trying to obtain some compensation from the company: "What are you going to do about my case? I have been home nine weeks now and have not received a penny—I think it's time that there was something from you as the National Health refuses to pay me anything. I am needing nourishment and the money, and I should have had 9 weeks wages now through no fault of my own."[19]

Nellie died in 1924, at the age of thirty-three, from respiratory failure. Her family brought a lawsuit against Turner Brothers Asbestos Company. At trial, Dr. William Cooke, a local pathologist, testified that Nellie's lungs had evidence of extensive fibrosis, and within this fibrosis he clearly saw "particles of mineral matter . . . of various shapes, but the large majority have sharp angles."[20] Comparing these particles to a sample of asbestos, Dr. Cooke made the obvious conclusion that the jagged fibers "originated from asbestos and were, beyond a reasonable doubt, the primary cause of the fibrosis of the lungs and therefore of death."[21]

Written up and published in the *British Medical Journal* in 1927, Nellie's case spurred an inquiry from the British parliament in 1930. The inquiry culminated in the paper "Occurrence of Pulmonary Fibrosis and Other Pulmonary Affections in Asbestos Workers," which was published in *The Journal of Industrial Hygiene*. The authors definitively identified a link between asbestos and pulmonary fibrosis, demonstrating that 66 percent of workers at the Rochdale factory where Nellie Kershaw worked were afflicted with pulmonary fibrosis.

In 1931, in response to this paper, the government of the United Kingdom established regulations to control dust exposure and provide compensation for those afflicted with pulmonary fibrosis from asbestos. This reaction prompted Thomas Legge, the former Chief Medical Inspector of Factories, to state in 1934: "Looking back in the light of present knowledge, it is impossible not to feel that opportunities for discovery and prevention of asbestos disease were badly missed."[22]

Despite Nellie's case and Legge's expression of regret, the mining and use of asbestos continued unabated worldwide for the next four decades, with few attempts at controlling dust exposure. A mining town sprang up in Western Australia, where beautiful but deadly blue asbestos was pulled out of the ground. Quebec, Canada, became the world's leader in asbestos production, from the appropriately named town of Asbestos. The United States, Russia, and Europe also dug large mines to harvest this miracle fiber.

The strength and resilience of asbestos are the same properties that make it so deadly. When inhaled, very little protective cough reflex occurs, and the fibers lodge themselves into the tissue of the lung. The body normally deploys cellular scavengers, like macrophages, to eat and digest foreign material and microbes, but the fibers of asbestos are too big and strong for the body's scavengers to digest. So they sit there, in the lung, forever.

With increased asbestos use, the prevalence of diseases such as lung cancer and pulmonary fibrosis increased, and a strange new cancer began appearing—mesothelioma. A cancer of the lining of the cavities of the body, typically either the lung or the abdomen, mesothelioma is one of the most lethal cancers, with a life expectancy measured in months. Unlike lung cancer, mesothelioma is not related to smoking, and with very few exceptions, its only known cause is asbestos. It grows like a python in the body, slowly encasing and suffocating organs. Chemotherapy is generally useless, with side effects worse than any benefit. Surgery may extend life a few months, but rarely more.

Studies have now shown that mesothelioma typically takes up to thirty or forty years after the exposure to manifest. It is mostly a workingman's disease, but it can also be a workingman's wife's disease, as women have been exposed from washing their husband's clothes. A handful of celebrities have died from mesothelioma, including the musician Warren Zevon, at age fifty-six, and the actor Steve McQueen, at age fifty. Another famous victim was Merlin Olsen, a fourteen-time Pro Bowl defensive player in the NFL, and later an actor on *Little House on the Prairie*. He grew up in rural Utah, and his exposure to asbestos was traced back to when he was a teenager in the 1950s working summer jobs at construction sites.[23] The use of asbestos has been drastically reduced in the United States. However, because of the long lag time from exposure to potential outcome, and the widespread use of asbestos in the 1970s, the incidence of mesothelioma has not yet declined, with on average about three thousand new cases per year.[24]

With gathering evidence of harm, bans on asbestos began in the early 1970s, and today the use of asbestos is banned completely in all twenty-eight European Union nations. The United States initially instituted a complete ban in 1989, which was overturned in 1991 by a circuit court, and the importation, processing, and distribution of products containing trace amounts of asbestos is still permitted. This includes goods such as disc brake pads and linings, gaskets, and roofing and fireproofing materials. Domestic production of asbestos ceased in 2002, but all told the United States imported on average about five hundred tons of asbestos per year between 2014 and 2018, mostly from Russia and Brazil.[25]

Russia still mines and uses asbestos on a large scale. The town Asbest, at the foot of the Ural Mountains, is home to the largest of the mines, and the dust there is insidious. As Tamar Biserova, a resident, commented in a July 2013 *New York Times* article, "When I work in the garden, I notice asbestos dust on my raspberries." So much dust blows against her windows, she said, that "before I leave in the morning, I have to sweep it out." In true Russian fashion, Nina Zubkova, another resident, wryly commented, "Of course asbestos dust covers our city. Why do you think the city is named Asbest?"[26]

Russia still mines about a million tons of asbestos every year, with hundreds of thousands of workers dependent on the asbestos industry for their livelihood.[27] Worldwide, some two million tons are mined every year. We are just a few years away from the hundred-year anniversary of Nellie Kershaw's death, and our lungs have still not learned how to digest this deadly fiber.

PART III

THE FUTURE:
THE LUNGS PROVIDE
A VISION OF WHAT'S
TO COME

Chapter 10

Curing the Incurable

There is a good deal we know about the lungs compared to twenty years ago, or even ten. Paradoxically, though, all the new knowledge is showing us not how far we've come, but how much further we have to go. The models we used in the past to represent what happens in the lungs have turned out to be far too simplistic, and as is typical, the designs of nature and the human body are proving to be far more complicated and mysterious than previously thought.

The systematic investigation of the lungs began over a millennium ago with the practice of dissection, which first arose in Alexandrian Greece in the third century BCE. After that, this practice was widely prohibited throughout the Western world and did not resume again until the early fourteenth century, in Italy. Over time, scientists became familiar with the gross appearance of the lungs, oblong in shape and deep pink in color. The left lung was noted to be slightly smaller than the right lung, to make room for the heart, and each lung was divided into several distinct lobes. They were spongy when touched, and clearly had airways leading into the gas exchange units, the alveoli.

With the use of the microscope in the seventeenth century, researchers discovered that the lung was comprised of several types of cells. Those along the trachea and bronchi looked very different from the ones deeper down, in the alveoli. The trachea and bronchi have squamous cells, which help protect tissue, as well as mucus-secreting cells to help defend against bacteria. The alveoli are lined with type I pneumocytes—long, thin cells that augment gas exchange—and type II pneumocytes, which secrete surfactant to help lubricate. With careful separation of the lung tissue, scientists were able to measure the total distance of all the airways in the human body, and it amounted to more than fifteen hundred miles, an incredible amount of tissue packed into so small an area. Even more striking was the number of alveoli in the average lung, some five hundred million in all.

With the advent of electron microscopy in the 1930s, magnifications of a million could be achieved, compared to two thousand with a regular microscope, and even more cell types were discovered within the lung. With the use of DNA analysis and genetic expression, even more have been revealed; in August 2018, researchers discovered a pulmonary cell called the *ionocyte*, which may transform our understanding of how the lung hydrates itself.[1]

With the discovery of each new cell, our picture of how these cells interact with the environment and each other becomes more complex. In addition to the cells that remain within the lung tissue, there are many others that come and go. In a process called *leukocyte trafficking*, for example, a variety of white blood cells mysteriously come in and out of the lungs in reaction to different bacteria, viruses, and physical insults. Some cells within the lungs have stem-cell properties and can differentiate into other cell types and rebuild the lung after injury, such as the airway basal cells of the bronchi, which help repopulate the lining of the lung after smoke inhalation, or the type II alveolar epithelial cells that repopulate the lining of the alveoli after other types of injuries. Understanding how all the cells in the lungs interact with one another in real time, each cell not static but expressing different proteins and hormones at different moments, is a monumental task.

Even as the picture of what keeps our lungs functioning gets stranger and at times more mysterious, moments of insight make us hopeful that we are beginning to understand the nature of the universe that is the lungs. Basic questions are being answered, such as how the lung gets injured, how it tries to repair itself, and what happens when this process goes wrong. These answers are accelerating progress on some previously incurable diseases.

*   *   *

Every morning for forty years, Dr. Gregory Halligan has driven himself to work at St. Christopher's Hospital for Children in Philadelphia. He is a pediatric oncologist, and he sees the saddest of the sad diseases. Only a special kind of doctor with a special kind of soul can handle kids with cancer, and Dr. Halligan is one of them.

The first sign of a change in Dr. Halligan's health came during late December 2013, when his energy level dropped. By February he was afflicted with shortness of breath, and before long he was gasping for air. Utilizing an oxygen meter at the hospital, he checked the oxygen saturation level in his blood. It was in the 70s. A normal level is at least 95 percent, and anything below 88 percent is considered dangerous. Dr. Halligan then did the wise thing and checked himself in to Hahnemann Hospital in downtown Philadelphia.

At about the same time, I got a page from the pulmonary fellow treating Dr. Halligan. "You need to come to the floor and look at this X-ray," he said. "I've never seen anything like this." The X-ray was waiting for me as I walked onto the ward, and it didn't take long to identify the problem—its calling card is unmistakable. Instead of a lot of air (which looks black on an X-ray) and a few spindly vessels in white, I saw a dense white infiltrate focused in both of Dr. Halligan's lower lobes. The lungs themselves appeared small, and so scarred they looked like rock. The disease, almost certainly idiopathic pulmonary fibrosis (IPF) or a close relative, had clearly been going on for a while. We flipped to the CT scan, and the more detailed chest X-ray confirmed what we saw earlier—massive amounts of scarring throughout both lungs.

In the consulting room, I introduced myself to Dr. Halligan and went over his history, asking when his symptoms had started and what else had been going on. He provided the exact dates of when his symptoms had started, and described how he had tried to help relieve them. We talked about what the disease could be, and what it couldn't be. I suggested starting a treatment regimen of steroids and antibiotics, as well as a biopsy, trying to sound hopeful while in the back of my mind I thought about the relentless nature of this disease, and the limits that exist in the practice of medicine.

Over the course of a week, Dr. Halligan seemed to get a little better. He was still on a lot of oxygen, but his exercise tolerance had improved, and he was sleeping well at night. We were able to back off a little on the steroids and decided to proceed with a lung biopsy. Keeping a patient on steroids for too long can decrease the yield of a biopsy by partially treating the disease, so our window to obtain a diagnostic piece of tissue was narrow. Dr. Halligan did well with the procedure, but the next morning when I went see him, he told me he had had a rough night. He had to cut short his morning shower, running out of breath after two minutes. Even after he put his oxygen mask back on, his breath didn't come back easily. Examining Dr. Halligan's lungs under the microscope in the pathology lab, we saw no inflammation, just sheets of fibrosis. What he had was not going away.

Dr. Halligan and I talked about the one thing that was left to do: I would call around town and see if he could be transferred to a center and worked up for an emergent lung transplant. At first Dr. Halligan was hesitant, but finally he agreed. The same was true of the transplant doctor with whom I spoke. Lung transplant centers want patients who are still well enough to walk into an office appointment, and with so many patients on the waiting list, lungs are a precious commodity. So I wasn't surprised when the lung transplant doctor suggested that I send Dr. Halligan home, promising to make an appointment for him in a week or two. I pushed back, insisting that he wouldn't make it at home, even with extra oxygen. The transplant doctor finally relented, though he emphasized that Dr. Halligan could not leapfrog anyone already on the list for transplant.

Two weeks after sending Dr. Halligan to the transplant hospital, I received an e-mail from the program coordinator saying he had been successfully evaluated and accepted for transplant. A few weeks later he got a call informing him that a new lung was being harvested for him. He got his transplant, and to the relief of many in the medical community he did well, and this story has a happy ending: A few months after receiving his new lung, Dr. Halligan was back at the hospital taking care of little kids with cancer.

* * *

Attempts to treat IPF are rarely successful: life expectancy with the disease is dismal, with a 50 percent survival rate of about four years.[2] It is a relentless disease that turns lungs to stone, one with no therapies that definitively increase life expectancy. This is why IPF is the most frustrating and disheartening disease in all of pulmonary medicine. Lung transplant is a salvage remedy; it does not treat the disease itself, and it comes with its own set of issues. Fortunately, our knowledge about the disease is improving, as is our understanding of what is happening when the cells of the lung start talking to each other during times of stress.

Dr. Paul W. Noble is the chairman of the department of medicine at Cedars-Sinai Medical Center in Los Angeles, and for three decades he has dedicated his career to understanding IPF. Dr. Noble first became interested in pulmonary fibrosis during his time as a resident at the University of California, San Francisco. A bone marrow transplant unit had been set up during his senior year, and many of the patients in the unit were experiencing respiratory failure. Death often ensued despite best efforts, and at autopsy all they found was fibrosis—patients' lungs had turned to stone.

Seeing these patients die so suddenly was a shock to Dr. Noble. It was also the start of the big question that would consume his entire career—what is the driving force for the lungs to create scar tissue? This interest carried forward into his fellowship at National Jewish Health in Colorado, which at the time had one of the biggest cohorts of pulmonary fibrosis patients in the country. At National Jewish Health, Dr. Noble saw typical patients with IPF, who often presented in an advanced stage, without much the physician could do.

Given the scope of the problem, Dr. Noble understood right away that this disease was going to be his whole career. Fibrosis patients represented the big mystery in all of pulmonary medicine, a disease so cryptic as to seem impenetrable. Dr. Noble was driven to at least try to treat the untreatable.

* * *

Descriptions of pulmonary fibrosis date back to ancient Greece, but because most people have vague symptoms, the disease went largely unrecognized for centuries, just another unknown cause of respiratory failure. This changed in the nineteenth century, as autopsies gained popularity as a way to learn about disease. In 1838, Irish physician Dominic Corrigan used the term *cirrhosis cystica pulmonum*, a reference to the diseased lungs' resemblance to a cirrhotic liver. In 1893, William Osler called the condition *chronic interstitial pneumonia* in his book *Principles and Practice of Medicine*. But no organized attempt to investigate pulmonary fibrosis occurred until 1944, when Drs. Louis Hamman and Arnold Rich, from Johns Hopkins, published an article describing four patients who all had died suddenly and were found postmortem to have dense fibrosis of the lungs.[3] A more formal classification system was set up in 1969, but over the years no progress was made on treatment. Patients were dosed with steroids, such as prednisone, or other immunosuppressive drugs, but it was suspected these had little, if any, impact on the course of the disease.

This was the state of the field when Dr. Noble started taking care of patients in the fibrotic lung disease program at the University of Colorado in the 1980s. For an anti-inflammatory medicine like prednisone to not work for a noninfectious disease was something new to the pulmonary community. At the time, inflammation was the prevailing theory for many diseases across medicine. The cells that initiate and propagate inflammation—those neutrophils and lymphocytes and macrophages that are made in bone marrow—migrate to an area based on a real or perceived threat and begin doing their work, for good or ill. They are the trigger for many diseases, but not apparently for IPF.[4]

By the 1990s, through rigorous study, it had become clear that medicines like prednisone were not an effective treatment for IPF, and lung doctors throughout the world realized that the entire field of idiopathic pulmonary fibrosis needed to be overhauled. The Pittsburgh International Lung Conference in 2002 began to move the needle forward. Alveoli are lined with two types of cells, alveolar type I and type II. The type I cell is the workhorse of the gas exchange unit, long and thin, facilitating the business of oxygen getting absorbed and carbon dioxide getting expelled. All told, 95 percent of the lining of the alveoli is made of type I cells. The type II cell is smaller, and cuboidal in shape. They produce surfactant, the lubricant that allows the lung to expand and contract. They are also the stem cells of the lung, able to differentiate into type I cells after injury. One further cell, the fibroblast, lives in the interstitium, the area between the alveoli and blood vessels. The fibroblast is the primary producer of collagen, the main ingredient of fibrosis.

In his summary paper after the Pittsburgh lung summit, Dr. Noble reported that the current science pointed toward a vast extermination of alveolar type II cells as the primary event inciting IPF.[5] Since there was no preceding inflammation, he posited that an injury in the form of something inhaled must cause toxicity to this cell type, which lives deep within the architecture of the lung. For some reason, likely the severity of the insult combined with genetic predisposition, type II alveolar cells not only die, but others fail to take their place. Since type II alveolar cells communicate with fibroblasts, moderating their production of collagen, their absence allows fibroblasts to produce increasing amounts of collagen, and fibrosis builds up. This likely happens slowly, over the course of many years, before a patient begins to notice any decline in function.

The obvious targets for therapy would be to protect or rescue the type II alveolar cells from death, or to prevent the fibroblasts from producing so much collagen. Prednisone and its relatives are very effective at stopping cells such as neutrophils and lymphocytes, but appear to have minimal effect on fibroblasts. The lung community joined with the pharmaceutical company Genentech to trial a unique medicine called Interferon Gamma, which had no FDA approved indications but had clear antifibrotic properties in the laboratory. That trial, published in the January 8, 2004, edition of the *New England Journal of Medicine*, was a watershed event.[6] The results, sadly, were negative, but the process was a huge step forward. A well-designed trial for pulmonary fibrosis had never been done before. Getting patients to participate in something that doesn't pay much and is a huge time commitment can be very difficult. It is also very expensive for the pharmaceutical company, costing about twenty million dollars just for the final phase 3 trial, to say nothing of phases 1 and 2. Getting a large, very expensive trial completed was a big feat, especially for a typically ignored disease like IPF.

Although officially the results were negative, the trial was very close to being a positive study. For a study to be considered statistically valid, the likelihood of the results not being from chance should be less than 5 percent. At 8 percent, just 3 percentage points short of that threshold, the results encouraged Dr. Noble and other researchers to think that they were on the right path targeting the fibroblast.

In 1972, a PhD chemist named Shreekrishna Gadekar submitted a patent for a drug that would change the course of IPF treatment. He named it AMR-69, and wrote in the patent that it had "excellent analgesic activity, marked anti-inflammatory activity and shows excellent anti-pyretic activity in test animals."[7] He also noted that "protection against noxious focal respiratory tract pathology (petechiae, edema, hemorrhage, focal infection, etc.) has been demonstrated in gross examination of rat lung tissues and microscopic examination of dog lung tissues following treatment with AMR-69."

Nothing happened with the drug until 1989, when scientist Samuel Margolin, founder and president of the Texas pharmaceutical company Marnac, obtained the rights to the drug to investigate it specifically for its antifibrotic activity. Margolin renamed it pirfenidone, and quite presciently stated in the updated patent that the goal of this new drug was to "prevent an excessive pathologic accumulation of collagenous scar or connective tissue in various body structures and organs." He specifically mentioned the lung.[8]

Studies on pirfenidone in the lung ensued in the 1990s, with the first trial, in 1995, entitled "Dietary intake of pirfenidone ameliorates lung fibrosis in hamsters."[9] Other studies followed, catching the eye of Intermune, a much bigger pharmaceutical company, which had enough resources to bring the drug to trial in humans.

The first success was a 2010 Japanese study, in which 275 patients with IPF were shown to have a slower lung function decline after taking pirfenidone than those who took a placebo.[10] This was followed up by the 2011 CAPACITY study, coauthored by Dr. Noble, in which 779 patients were also shown to have a more modest and slower lung decline than placebo patients.[11] In concluding the study, Dr. Noble and his colleagues stated that pirfenidone was "an appropriate treatment option" for IPF patients. This sentence held massive significance: it simply had never been written before. However, the FDA disagreed with Dr. Noble's conclusion. In the agency's estimation, the effect was modest at best: statistically but not clinically significant. The FDA wanted another, bigger trial.

The fourth trial of pirfenidone in IPF, again coauthored by Dr. Noble, was published in October 2014 in the *New England Journal of Medicine*.[12] The study on another drug, nintedanib, which also works on an antifibrotic pathway, was published in the same issue.[13] To the joy of many clinicians and patients, as well as Dr. Noble and his colleagues, pirfenidone and nintedanib showed a modest but real benefit, and both were subsequently approved by the FDA. The drugs are not without controversy, costing about $100,000 per year for very modest improvements, which come with potential side effects. But they are the first to make even the slightest step forward in the fight against IPF.

Paul Noble likens drugs such as pirfenidone to weed killers—they can clean things up, but they don't pull the disease out by the root. Sooner or later it will return. Ultimately, Dr. Noble believes we need two different drugs to really help patients with IPF—the first a drug like pirfenidone to target the fibroblast, the second to restore the type II alveolar cells whose initial death likely triggers uncontrolled growth of the fibroblast.

The answer may lie, as documented in years of painstaking experiments by Dr. Noble, in a gel-like substance that is found all over the body, a substance most pulmonary doctors don't even know is in the lung. Called hyaluronan, it serves as a protective coating in various places in the body, notably in joint spaces, where it cushions the impact of wear and tear. With its excellent viscoelasticity and high moisture retention capacity, it is present in significant quantities in the skin and eyes. It is also produced in the type II alveolar cells of the lung, where it forms a protective framework in the alveoli.

Dr. Noble has been interested in hyaluronan for a long time, and he has published a series of experiments in the high-profile journal *Nature Medicine* documenting its critical role in lung health and in repair in pulmonary fibrosis.[14,15] Using a mouse model, he showed that impairing the ability of type II epithelial cells to appropriately interact with hyaluronan led to cell death and increased lung injury. He also observed that adding extra hyaluronan to a lung matrix protected it from damage from toxic inhalants. Further experiments showed that alveolar type II cells in patients with IPF had reduced surface hyaluronan, along with reduced ability to regenerate, even when placed in an environment that was conducive to lung-cell growth.

This line of inquiry is hugely promising. Just producing this type of data, when twenty years ago lung doctors were blindly throwing prednisone at patients with IPF, is a colossal step forward. Whether hyaluronan is going to be to IPF in the twenty-first century what surfactant was to neonatal distress syndrome in the twentieth century is not yet clear, but its potential appears significant, and Paul Noble's goal is to develop a drug to increase hyaluronan in the lung so it can do its job of protecting the type II alveolar cell.

Bill Vick doesn't remember a specific day when his breathing started to decline. It was more of a slow dropping off, during which he kept telling himself everything was fine. But in retrospect, he knows that deep down he realized something was wrong. Although he was seventy-two years old, his fitness and health had been exceptional. Finally, though, the nag of breathlessness and a cough were impossible to ignore, and he saw his primary care doctor, who sent him to a specialist. A lung biopsy led to the final diagnosis in September 2011, the choking amount of fibrosis obvious under the microscope.[16]

Bill's description of IPF is the best one I've ever heard. He calls it the "ninja disease"—it sneaks up on you without warning and takes you out. There is nothing you can do to recognize it is coming, and no treatment yet exists to get rid of it once it's there, outside of a lung transplant. Bill went through the usual stages of reaction to an unforeseen and deadly diagnosis, from anger to pain to sadness to acceptance, but he retained some anger at the medical community for knowing so little about the disease, and for not doing enough to let the public know about it. In the United States, more than two hundred thousand people live with IPF. Forty thousand die every year. This is the same number of people who die from breast cancer annually, except there are no walks or pink ribbons or huge foundations for IPF. IPF is a ninja disease that few people talk about.

Bill has taken it upon himself to lift IPF out of obscurity. He formed the group PF Warrior to address challenges IPF patients face, offering advice and comfort to other sufferers. Upon meeting a newly diagnosed man who was clearly in the sadness and depression phase of his reaction, Bill assured him that his own life was as full as it had ever been. He was fighting the disease by working out every day, enjoying every moment with his family and grandchildren, and continuing to work full time—in short, he was living his life, even as his lungs and breathing declined. PF Warrior has expanded to seventy-five members, who regularly travel to meet with newly diagnosed patients, providing hope while research on the disease continues.

To move this research forward, we will have to continue to improve our models of IPF, expanding them out to reflect our growing understanding of the lung environment. The goal will be not just to analyze the lung as it exists in a particular moment in time, but to create real-time three-dimensional systems to examine how the many different cell types interact with each other under different circumstances.[17] The interactions are intricate and complex, the variety of different genes turned on in different cells under different circumstances almost mind-boggling. There is a world that we are just beginning to appreciate, but we are finally beginning to glimpse what can and will happen under different circumstances. With this understanding, it is not unreasonable to think that an effective drug that helps reverse fibrosis in the lung is not far off.

Chapter 11

Getting Personal with the Lungs

The lung is a wonder of cellular movement and cooperation. Lung cells participate in an elaborate system of communication in order to carry out their twofold mission: to keep air moving freely through the branching network of tubes while keeping noxious material out of that network. The lung's complex defenses consists of goblet cells, which secrete mucus to trap invaders, and ciliated cells, topped with tiny hairs that beat in unison to drive the mucus-coated intruders out. Unwanted dust, bacteria, and viruses are removed from the lungs on a kind of conveyor belt, designed for the removal of debris, that carries unwanted material back into the atmosphere, where it is generally harmless.

This whole defense system, known as the *mucociliary escalator*, does an enormous amount of work, and all this work produces a lot of "cellular turnover," that is, cell death and regeneration. Cellular turnover occurs in all organs, but the rate of turnover varies widely throughout the body. On one end are the organs with very high turnover rates, such as the bone marrow, breast, and gastrointestinal tract, while at the other end are the organs with very slow turnover rates, most notably the brain and heart. The lungs sit exactly in the middle of this scale, with a moderate rate of cellular death and regeneration.

Normally, higher rates of cellular turnover lead to greater potential for error, increasing the likelihood a cell will mutate. This is why tumors of the breast, colon, and bone marrow are quite common, while those of the heart and brain are less so. Historically, this pattern has not held for the lungs, which, prior to the twentieth century, were exceptionally resistant to developing cancers despite having a moderate rate of cellular turnover.

This, unfortunately, began to change at the beginning of the twentieth century, and one story drives this point home. In 1919, future thoracic surgeon Alton Ochsner was a medical student at Washington University in St. Louis. He had a love of the heart and lungs, and so when the entire class was invited to see the autopsy of a death from a rare lung disease, he jumped at the opportunity. The chief pathologist, Dr. George Dock, assured the class that it would be worth their time; this was a disease they would likely never see again. That day they would be doing an autopsy on somebody who had died from lung cancer.[1]

Sadly, as we know, this has changed. Today, lung cancer kills more people each year than the next three deadliest cancers combined. We know the cause well—with the advent of widespread cigarette smoking we have changed the lung from something almost immortal into something quite mortal. Some one hundred years into this malignant experiment, we must now put an end to it. Fortunately, there are signs we can, especially with our advancing knowledge of personalized medicine.

Late one afternoon in 2014, my team received a phone call from the primary medical team taking care of Glenda Abney: she was a middle-aged African American woman who presented to the hospital with pain in her chest, which proved to be caused by a large mass in her lung. I sat down at a computer with the fellow and looked at Glenda's X-rays. What I saw was not pretty—a big, jagged mass of white in the middle of the lung. A jingle ran through my head—*the rumor is tumor, the issue is tissue*—meaning, we think it's cancer, but we need a biopsy. A third verse completes the rhyme in most cases: *And the answer is cancer.*

Even though I'd done this exact same consult hundreds of times, I rehearsed what I was going to say in my mind before entering the room. I told Glenda she had a spot on her lung and she needed a biopsy. Her reaction, like all reactions, was unpredictable. For Glenda, it was not one of panic, or a million questions, or even a moment of regret. She just looked up at me, her eyes firm and resolute, and said, "Well, doc, we gotta do what we gotta do. So let's get it done."

I left the room and went to document our conversation in the chart. As I wrote, the frustration and anger that comes over all lung doctors faced with an almost certain cancer diagnosis came over me. We have failed these people on a very deep level. And almost always, because of the nature of lung cancer, there is no easy solution.

A day after I had the conversation with Glenda about the mass in her lung, I stood in the surgical suite watching my fellow numb-up Glenda's nose and throat with lidocaine so we could stick a camera into her lungs and get our biopsies. Glenda's demeanor hadn't changed in the twenty-four hours since I had met her. She still answered my questions in succinct sentences, not revealing any emotion behind them. This was unusual to me. Most patients facing a cancer diagnosis exhibit some level of fear, or anger, or distrust. But not Glenda.

The thin camera went into her nose, back into her throat, and then hovered over her vocal cords. We sprayed some lidocaine over them, then dipped the camera through her trachea and deeper into her lungs. The camera cannot go very far, and what we could see in the large airways looked unremarkable. I had the fellow put the camera in the area where we thought the cancer was, based on the CT scan, then we turned on the X-ray machine to illuminate her lung. There, in an instant, the mass sprang into view, a sinister, white, jagged ghost within the dark lung. We got out our biopsy forceps and threaded them out into the deeper part of the lung, where the mass could be seen on the X-ray machine, and pulled some tissue out.

I saw Glenda back in the office a few days later. Sitting in a chair, hands calmly folded in her lap, she patiently awaited my arrival. I wasted no time telling her what the pathologist had already told me—she had lung cancer. She took the news exactly the way I thought she would, the exact opposite of the way most patients do: "Well, I think I've got a few more years left, so let's go ahead and do what we gotta do, doc."

I told her she would need a PET scan to look for any other disease in her body. I told her I didn't know if she was a surgical candidate, but we were going to explore every option.

"That's about what I thought, doc," she replied.

With those words she summarized what I believe most lung cancer patients are thinking when they get diagnosed. They have been smoking a long time, and the risk to their lungs is no secret. Glenda also clearly understood this. We got her the best care available, and she did well initially, but eventually, like most lung cancer patients, she passed from her disease, and way too young.

Dr. George Dock's 1919 prophecy that Alton Ochsner wouldn't see another case of lung cancer proved accurate until 1936, when then–Dr. Ochsner saw his next case. Shortly thereafter he saw another, and then another—nine cases in six months. Most of them were World War I veterans, products of the era during which the mass marketing of cigarettes to soldiers took hold. With the power to alleviate stress and boredom, cigarettes came to be seen as part of America's arsenal during World War I, prompting American General John Pershing to state in 1917, "You ask me what we need to win this war. I answer tobacco, as much as bullets."[2]

Dr. Ochsner quickly made the connection between smoking and lung cancer, writing presciently to a colleague in 1939: "In our opinion the increase in smoke with the universal custom of inhaling is probably a responsible factor [in the increase in pulmonary carcinoma], as the inhaled smoke, constantly repeated over a long period of time, undoubtedly is a source of chronic irritation to the bronchial mucosa."[3]

Between 1930 and 1964, the voices on the side of science and medicine were loud and clear regarding the possible connection between smoking and lung cancer, but the voices on the side of the tobacco industry were just as loud in opposition. It was an ongoing debate in the minds of most Americans, which was finally resolved by two experts from London, physician Richard Doll and statistician Austin Hill. Dr. Doll was a smoker himself, and along with Austin Hill in the late 1940s, he was given the task of trying to understand the extraordinary increase in lung cancer cases seen in London's hospitals during the preceding twenty years.

Some blamed tobacco smoke, others pollution from cars, dust from the new asphalt roads, and smoke from industrial plants. Previous studies had been published tying smoking to lung cancer, but none had had enough statistical rigor to hold up to scrutiny. In order to definitively settle this debate, Dr. Doll and Austin Hill employed a case-control study, a simple but powerful analysis that helps define risk. They were one of the first to use it, and its impact on our thinking about the effects of tobacco was profound.

They set up their study by asking twenty hospitals in London and the surrounding area to report any case of lung cancer admitted to the hospital. Their group would then interview the patients about prior occupations, exposures, and habits, including smoking. But then—and this is what set their study apart—they would put those same questions about exposure to age- and gender-matched control subjects admitted to the hospital at the same time for other reasons. With this study, Doll and Hill moved the analysis of lung cancer out of the realm of the observational and into the realm of rigorous epidemiology.

Between April 1948 and October 1949, their team interviewed 709 patients with lung cancer, and then 709 patients without lung cancer who served as controls. They then created a table comparing the gender ratio within the groups, along with comparisons of their age, place of residence, and socioeconomic status. All of this was an attempt to control for the elements of observation that can happen by chance.

The results Doll and Hill reported in their 1950 paper were conspicuous.[4] A few of the patients were lost to incomplete follow-up, leaving 649 subjects in each group. In both men and women, the difference between those who had lung cancer and those who did not was the consumption of cigarettes. Everything else was similar—their age, their exposure to road dust, industrial plant smoke, and pollution from cars. The only thing that was divergent was the smoking, and using statistics, Doll and Hill proved this was not because of chance. They also showed an increased incidence based on the total number of cigarettes the patient had smoked in the past, again bolstering their hypothesis that exposure (smoking) leads to disease (lung cancer).

Word of Doll and Hill's findings spread around the world, including to the tobacco industry. Interestingly, the tobacco companies did not attack the science themselves, understanding that science and physicians were held in high regard. Instead, they hired pro-tobacco physicians to put out statements refuting claims of a correlation, and they formed the Tobacco Industry Research Committee, in December 1953, to begin publishing (selective) studies of their own. A decades-long battle ensued between science, the government, and the tobacco industry, until the truth that Doll and Hill had uncovered would be completely accepted.

In the nineteenth century, the famous German pathologist Rudolf Virchow articulated the Latin phrase *omnis cellula e cellula*, or "all cells from cells." This dictum certainly holds true for cancers, which start out as a single cell that both lives longer and divides faster than the other cells in the body. The offspring of this cell also inherit these belligerent traits, and over time they crowd out the normally functioning cells of the organ they originated in, and then often jump into the bloodstream to seed another organ, in a process known as metastasis.

Lung cancer is one of the most lethal diseases because of how rapidly these cancer cells can grow and divide. Within the world of cancer, some are much more aggressive than others. In regards to lung cancer, the cells are on the biological equivalent of steroids, angrily dividing and growing in rapid fashion. Because of this, 80 percent of patients initially come to their physician with stage III or IV disease, both of which are very advanced.[5] The average five-year survival rate for a new lung cancer diagnosis is about 18 percent, significantly lower than breast cancer at 90 percent and even colorectal cancer at 65 percent. Even for those with early-stage lung cancer, which is operable, the survival rate at five years is about 56 percent.[6]

In contrast to lung cancer, the cells of colon cancer grow slowly, and it can be effectively monitored. At first, it's a premalignant lesion, then a small local cancer, then it moves to local lymph nodes. A colonic lesion takes a predictable amount of time to grow, and at each point there is an opportunity for surveillance and intervention. Get your colonoscopy every ten years—for those at high risk, every five years. Lung cancer is the opposite. Over half of lung cancer patients present with metastatic disease. One year one's chest X-ray can be completely clear, and then twelve months later a doctor is explaining that the pain in one's chest is from a mass in the lung and the cancer may have metastasized to one's spine and brain.

This biology makes lung cancer by far the leading cause of cancer death for both men and women in the United States, with about 160,000 deaths per year, far outpacing the 40,000 breast cancer deaths every year and the 28,000 deaths from prostate cancer. Lung cancer causes more deaths per year than colon, breast, and prostate cancer combined.[7]

We are, as a species, influenced by emotion, so it's understandable that the ratio of scientists to patients with pediatric leukemia is very high compared to the ratio for other diseases. Since lung cancer is the number one cancer cause of death, some correlation with the level of federal funding for research would be expected. In fact, however, lung cancer receives about one-half the funding of breast cancer, although it has a four-times higher death rate.[8] Indeed, lung cancer is the stepchild of cancers, the least talked about and least recognized, with the most patients suffering.

Different ethnic groups also suffer more than others. A major study, published in the *New England Journal of Medicine* in 2006, compared lung cancer rates in African Americans, Native Hawaiians, Latinos, Caucasians, and Japanese Americans.[9] Among those who smoked twenty or fewer cigarettes per day, which is about 80 percent of smokers, African Americans and Native Hawaiians were twice as likely to develop lung cancer as Caucasians, and three times as likely as Latinos or Japanese Americans. With higher levels of smoking, those differences started to disappear.

The reasons for these ethnic differences are not clear. They could not be explained by any obvious differences in the patients' diets, occupations, or socioeconomic status, which leaves some potential genetic explanation, and/or some theories of different metabolism of carcinogens. Compared to other ethnicities, for example, African Americans appear to retain higher levels of nicotine in their bodies after smoking the same number of cigarettes, though no definitive conclusions have been drawn from this.

Happily, since the 1960s the rate of smoking has dropped at a very fast rate among African Americans. In 1960, about 60 percent of African American males smoked; today about 17 percent smoke. African American women have a lower rate of smoking today compared to either Caucasian men or women, at about 13 percent. The rate of smoking among African Americans aged twelve to seventeen is less than half the rate for Caucasians in that age group, at 3.2 percent.[10]

Overall cigarette smoking rates have declined in the United States from over 40 percent in the 1960s to an almost unprecedented 13.7 percent in 2018.[11] Still, over 30 million Americans use cigarettes. An increase in e-cigarette usage, cigars, and hookah pipes is also slowing the overall decline in tobacco use, especially in teenagers, and cases of bad outcomes due to vaping are piling up. The battle is, unfortunately, far from over.

In his 1971 State of the Union address, President Richard Nixon declared war on cancer, and later that year he signed the National Cancer Act, which stated that the country needed to "attempt to find a cure, and to make a total national commitment to defeating this disease."[12] While aware of the many different kinds of cancers, Nixon talked about cancer as a single disease. Since then we have come to focus on organ-specific cancers, and even more recently on individual cancers, each having unique characteristics that can potentially be targeted for therapy. This was reflected in President Barack Obama's 2015 State of the Union address, in which he specifically addressed the concept of personalized medicine, recognizing each person's cancer as a unique disease.

James "Rocky" Lagno stands as a testament to this new approach to cancer treatment. In November 2010, visiting family for Thanksgiving, he began to notice a cough that wouldn't go away. He had rarely been sick before. His local primary care doctor diagnosed him with pneumonia and gave him an antibiotic. Nobody gave it another thought, because there was no reason to. Rocky didn't smoke, he drank only socially, and he didn't take any medications.

But the cough continued, and a few months after experiencing his initial symptoms he hacked up some bright-red blood. He drove himself immediately to a local urgent care center and got a chest X-ray. Since it showed a lung mass, he was then given a CT scan and a biopsy. A few days later, with his wife Geralynn at his side, Rocky was face to face with an oncologist, who acknowledged he had nonsurgical, advanced disease. The oncologist told Rocky that there was hope, and that this was treatable, but with the tyranny of the statistics weighing on him, he also told Rocky to make a "bucket list" of things he wanted to do, with the implication that he should start working on them sooner rather than later.

The oncologist then mentioned to Rocky and Geralynn some clinical trials of new medications that were showing dramatic results in patients with lung cancers that exhibited certain genetic changes. These genetic changes were more likely to occur in patients like Rocky, who never smoked. From her own research, Geralynn learned that tumors can look identical under the microscope but have formed through different genetic pathways.

Individual lung cancers have specific genetic mutations that drive cellular growth. With many cancers, if you shut off one switch, other circuits take over and the cancer continues to grow. Some cancers, though, have a main power switch, and once you discover what it is, you can shut it off with a targeted medicine, causing cells to stop dividing and effectively killing the cancer. If Rocky's cancer had one of two particular mutations—the anaplastic lymphoma kinase (ALK) gene or endothelial growth factor receptor (EGFR) protein—that mutation could be exploited as a "kill switch." However, only about 4 percent of lung cancers have these genetic changes.

Strangely, the oncologist they saw at their second consultation had little interest in taking another biopsy of Rocky's cancer to see if it harbored one of the two genetic changes. Rocky soon commenced traditional radiation and chemotherapy, which kills cells en masse, some of them cancer cells, some of them not. The usual things happened to him, such as loss of energy and hair. With over thirty straight days of radiation to his chest, the skin on Rocky's back bubbled up and bled, excruciatingly painful to the touch. Even worse, the radiation did not work. Geralynn asked again to get his tumor tested for specific mutations, and the doctor finally relented. And indeed, Rocky had one of the two mutations against which a targeted medicine could be used.

Rocky and Geralynn, not surprisingly, decided it was time to change physicians. They found their way to Dr. Alice Shaw, a rare oncologist who has chosen to specialize in lung cancer. She works at Massachusetts General Hospital, where she is a "translational researcher," one who straddles the border between the two worlds of science and patient care. She quickly put Rocky on crizotinib, a medicine that specifically targets the ALK mutation that was present on his tumor. What happened next was nothing short of extraordinary. Rocky noticed an immediate improvement in his well-being and energy level, and the X-rays backed him up. His tumor was shrinking for the first time.

Over the next several years, Rocky would go through an almost continual roller coaster of treatments and complications. Fluid around his heart and lungs required tubes and drainage. A new focus of cancer appeared in his lung, followed by metastases in his brain. He went through radiation to his brain and began traditional chemotherapy again. An irregular heart rhythm increased his rate to 130 beats per minute, and blood clots required him to start a powerful blood thinner. Even worse, the crizotinib stopped working. Newer medicines that targeted the ALK mutation were in clinical development, but Rocky wasn't eligible because of multiple complications. Undeterred, Dr. Shaw successfully lobbied the drug company for compassionate use. Rocky and Geralynn's mantra was put to the test: "tough times don't last, tough people do!"

Through all of the complications and advancing disease, Rocky and Geralynn spoke freely and openly about his cancer diagnosis, posting blog entries privately for friends and family to read. July 28, 2016, was a significant day for Rocky, and he wrote one of his longest blog posts ever: "We stand here today, exactly five years from a terrible prognosis with my health generally good, my spirits high and living a quality of life heretofore unthinkable for stage IV cancer patients. As with many cancer patients, I wanted to have it remain quiet but what a mistake that would have been. There were some very difficult times early on and also some tough after effects as time went on. Radiation burns, blood clots, fluid on the heart and lung with subsequent surgeries. Through it all I was able to stay positive and regain strength due to my wonderful wife's support and care."

In the spring of 2019, Rocky was on his fourth ALK inhibitor, and able to stay just ahead of the cancer. He spoke then of a side effect of vivid dreams from the newest medication, which he described not with frustration or anger, but with a laugh, an acknowledgment that this was just another thing he would deal with. Sadly, he passed away in November 2019, but his survival for nine years after his initial diagnosis of stage IV lung cancer was an unheard-of outcome until recently.

The future of lung cancer treatment lies in continued genetic analysis of tumors, with the expectation that targeted therapy can be increased from the current level of 4 percent of patients. Other approaches are also being investigated, one of the most interesting of which harnesses our own immune system to attack the tumor. The medicine pembrolizumab was recently approved by the FDA for lung cancer. Its mechanism is unique, as it works by attaching to lymphocytes and activating them to attack cancer cells. A study showed that in people with very advanced disease, those who received pembrolizumab were 70 percent likely to be alive after one year versus 50 percent for those who received only standard chemotherapy.[13]

In addition to cutting down on smoking rates, we need to determine what else we are inhaling that is causing lung cancer. Lung cancer patients like Rocky, who never smoked, are becoming more common. The data on what these cancer-causing inhalants could be is scarce; the field is in its nascent stage. Radon gas appears to be a contributor. A naturally odorless gas, it occurs in harmless amounts outdoors but can build up inside homes constructed on soil contaminated by naturally occurring uranium deposits. Secondhand smoke is also a contributor, as is air pollution. Why some people do and others don't develop cancers at similar exposures goes back to their genetics, of which there is much more to be understood. The lungs have traditionally had a robust system of cell death and replacement. We must try to understand what has changed in our environment that is altering this balance.

Chapter 12

The Breath and the Voice

We breathe, most importantly, to bring oxygen from the atmosphere into our bodies and to release carbon dioxide in return. Another crucial aspect of the breath, and one that is rarely discussed, is that it gives us our voice. Breathing generates speech, a fundamental element of our experience as human beings. Speech comes to us organically as children, and significantly impacts our daily interactions. An obvious extension of speaking is singing, and the natural instrument we carry around with us has influenced our culture as profoundly as speech.

The production of sound is something we take for granted, but it involves a complex network of interactions, and the crowded area in which the voice box lives is fraught with potential problems. The intake of air begins with the nose or mouth. The mouth, though, functions as more than just an airway; it's equally a part of the digestive system. Extending down from the mouth is the pharynx, the back of the throat, which splits in two: In the front it becomes the windpipe, or trachea, and in the back it becomes the esophagus. At the juncture of this split is the larynx, or voice box, which sits right on top of the trachea. Inside the voice box lie the vocal cords, opening and closing as air from the lungs flows past them to produce sound.

The vocal cords themselves are two thin white strips of tendon, one to two centimeters in length and a few millimeters wide, that are stretched tightly over the top of the trachea to maximize the effect of airflow. In the front they touch, while in the back they are separate. Their movement to produce sound is akin to that of wiper blades on a car, coming together to just the right distance for every single syllable

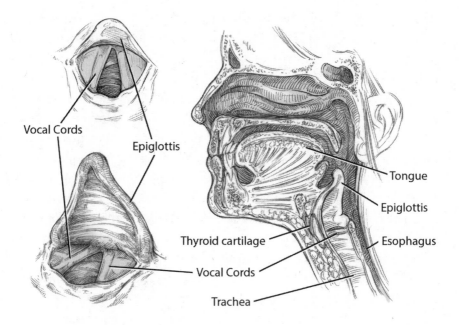

The vocal cords, on the left from above, on the right from the side.

we produce. Considering how much they are used over a lifetime, the amount of work they do for their size is remarkable.

Voice and song are produced when air expelled from the lungs rushes past the vocal cords and causes a vibration. That vibration creates phonation, or sound, the physics of which are complex. In short, air pressure builds up from the lungs beneath the vocal cords, pushing them slightly apart and causing them to oscillate. This oscillation has the effect of chopping up the air flow into brief puffs, emitting waves of energy, which the ear picks up as sound. The amplitude and frequency of the waves produced from the chopped-up air is determined by how far apart the vocal cords are and the natural frequency at which they vibrate. On average, men have longer vocal cords than women, at seventeen to twenty-three millimeters long, and the frequency of their sound wave is about 125 hertz (Hz, defined as the number of wave cycles per second). Women's vocal cords are about thirteen to seventeen millimeters long, with a 210 Hz frequency, producing a higher-pitched sound, while children's voices are even higher, with a vocal-cord beat frequency producing sound waves of about 300 Hz.

The vocal cords are supported by a series of muscles that augment movement, and cartilage that protects them from damage (the vocal cords, muscles, and cartilage together make up the voice box). One of the protective structures is the thyroid cartilage, or Adam's apple (women also have Adam's apples; men's are just more prominent), which covers the front portion of the voice box. Another, the cricoid cartilage, forms a ring around the trachea. If you feel the thyroid cartilage in your neck and then slowly palpate down the midline, you will encounter a small depression before you reach the other bony structure, the cricoid cartilage. This is the cricothyroid space, where an emergent breathing hole can be bored (a cricothyrotomy) if food clogs the trachea.

Another important structure in this area is the epiglottis. A triangle-shaped piece of tissue that sits at the base of the tongue, it stands guard from above by collapsing over the voice box with every swallow to prevent food or drink from going down the trachea and into the lungs. Air usually moves in and out of the larynx, flowing past the vocal cords with each breath. But the process must pause when anything is consumed so the epiglottis can snap into action, falling over the vocal cords to ensure food and drink enter only one pipe—the esophagus.

We may think of the vocal cords as being involved only with speech, but they also play an important role in swallowing, breathing, and many other daily functions. Surprisingly, they play a role in activities such as pushing a car, lifting a heavy weight, going to the bathroom, and even giving birth, since the pressure in our chest that helps us produce the force to push couldn't happen unless our vocal cords closed appropriately.

Many things can go wrong in this area. Should the epiglottis fail to collapse properly, as can happen in the elderly or those with a brain injury, food or liquid may cascade into the lungs from above. Termed aspiration, this process can trigger a severe cough, as well as a lung infection if the aspirated amount is significant. Acid refluxed from the stomach and mucus expectorated from the lungs also have the potential to injure the vocal cords. For breathing, if one of the cords does not stay open at rest, as it should, the flow of air will be compromised. If both cords are damaged, suffocation can occur.

Upper-respiratory anatomy varies widely in nature. Unlike mammals, with our shared respiratory-digestive passageways, fish have separate areas for breathing (gills) and eating (mouth). With these anatomically discrete locations, fish can eat and breathe simultaneously, which humans of course cannot. Birds also have a unique system compared to mammals. Quite remarkably, if you open a bird's mouth to examine it, a large hollow tube, the bird's trachea, positioned high in the throat, looks back at you.

The voice box of a bird lies deep in its chest, at the base of the trachea instead of at the apex. It sits at a level equivalent to that of the human breastbone or sternum. Birds don't have vocal cords, but instead have a syrinx, a series of muscles in the walls of the trachea that constrict it to produce sound. This is a much better protected space for the voice box than the one we have, allowing a freer flow of air and enabling the harmonious calls, songs, and whistling with which we are all familiar.

The importance of the voice is summed up by the quote ascribed to Danish linguist Otto Jespersen: "In the beginning was the voice. Voice is sounding breath, the audible sign of life."[1] Our voice allows us to convey not just our thoughts, but our mood, our emotions, and our intentions. Our voice is something deeply personal, influenced by family, friends, and where we grew up, a personal calling card that also serves as a gateway between our thoughts and the outside world.

The value of the voice, though, becomes clearest when it is lost— especially when that loss is compounded by a loss of breath, which it often is. This is a frustrating medical condition not only because of what it does to the life of the voiceless, breathless person, but because of the scarcity of medical specialists who know how to effectively deal with the problem. Dr. Claudio Milstein, at the Cleveland Clinic in Ohio, is trying to improve this situation. His approach teaches us not only how voice and laryngeal disorders should be diagnosed, but also how medicine should be practiced.

The Cleveland Clinic was founded in 1921 by four physicians who came together to form a practice with the motto "to act as a unit." Three of them—George Crile, Frank Bunts, and William Lower—were surgeons who had served together at the Lakeside army hospital unit in France during World War I. While there, they saw professionals from different specialties—surgeons, internists, and nurses—work happily and efficiently together while focusing on their mission to help wounded soldiers.[2]

The three surgeons recruited internist Dr. John Phillips, also a former WWI military physician, to carry forward the principles of collaboration and discipline while working toward the goal of delivering effective healthcare. Dr. Crile, the de facto leader of the group, wrote that "mediocrity well organized is more efficient than brilliancy combined with strife and discord."[3] This group practice—physicians working collaboratively and supporting each other—embodied a founding principle and model that endures today.

The Cleveland Clinic opened on February 29, 1921, as a strictly nonprofit hospital focused on patient care, education, and research. Dr. William J. Mayo, a founding member of the Mayo Clinic, in Rochester, Minnesota, spoke at the opening ceremony, highlighting the importance of working together in medicine: "Properly considered, group medicine is not a financial arrangement . . . but a scientific cooperation for the welfare of the sick."[4]

Dr. Claudio Milstein embodies George Crile's idea of the perfect doctor: highly competent, collaborative, and committed to education and research. A speech scientist at the Head and Neck Institute at the Cleveland Clinic, he is currently the director of the Cleveland Clinic Voice Center. He is not a lung doctor, a lung surgeon, or a lung researcher, but his area of expertise is closely related to pulmonology, and more importantly, his work is essential to patients with breathing problems.

Dr. Milstein was initially interested in theater and became a stage director. In doing this work, he gained an appreciation of the importance of the voice and its complexity. After attending medical school and obtaining a PhD in speech sciences, he decided to dedicate his professional life to people affected by dysfunction of the voice box. Dealing with the voice box is delicate work, because a physician who alters the larynx to help with speech must be careful not to compromise a patient's swallowing or breathing. Dr. Milstein deals with complicated cases, and he commonly sees people who have had no voice for months or even years.

Kevin Neff had long searched in vain for answers to his voice problems, which began in 2007 when, at age forty, he picked up a virus that spread to his larynx. It is common for a virus to spread to the larynx, but Kevin's case was unusual. His voice not only became raspy and weak, it disappeared entirely after three weeks and then refused to come back. Not surprisingly, he lost his job in sales. He saw physician after physician, only to be given the throwaway diagnosis of "functional dysphonia," a puzzling condition in which the patient loses their voice while the vocal cords and the muscles around them all appear normal.[5]

Kevin tried lip reading, writing on a white board, and briefly using sign language, but none of these substitutes enabled him to communicate effectively. Finally, an iPad with a text-to-speech application helped him rejoin the world. But he still had no voice of his own.

Despite his challenge, Kevin enrolled at Iowa State University to study finance, learning to see the world from a disabled person's perspective. He discovered that people are not always patient with those who have handicaps, and in conversation many people are more focused on what they are going to say next than on truly engaging. In response, Kevin learned to listen intently, to reply using precise language on his iPad, and to ask for help whenever it was needed. He came to terms with his condition, believing that he had become disabled for a reason and that his struggles had made him a stronger and better person. But when his father-in-law sent him an article about a woman who had recovered her voice after a single visit to the Cleveland Clinic Voice Center, he made an appointment.

Dr. Milstein determined that the muscles of Kevin's voice box had constricted and were now stuck, unable to contract. The virus had caused inflammation, and then scar tissue had formed, paralyzing the muscles. To combat this problem, Dr. Milstein began to manually massage Kevin's throat, eliciting loud snaps, crackles, and pops as the fibrous scar tissue began breaking up. This allowed the muscles to start contracting normally, allowing them to do their job of moving the vocal cords around to elicit sound. Remarkably, Kevin began voicing single letters, then strung together an *m* and an *e* to form his first word in almost four years. Over the next few weeks, with more exercises, he was able to build his voice back up to normal.

Kevin would go on to graduate summa cum laude from Iowa State University, where he gave one of the graduation addresses. He started the address using his old computer-assisted iPad voice, which many people in the audience recognized as his "normal" speech. He then shifted dramatically to communicating with his own voice, now full and deep, and told of his journey of adjustment, learning, and then recovery.

Not all voice box problems manifest as Kevin's did, and very often breathlessness is part of the problem as well. Laurie Danders, a resident of Irons, Michigan, lived in fear and isolation for several years, saddled with an illness that no physician could diagnose. Occasionally, and quite suddenly, she would lose her breath and, along with it, her voice. It was a terrifying experience. Her brain would clamp down, and panic would cause her body to be overwhelmed with stress hormones.

Laurie never knew how long these episodes were going to last. She would feel spasms in her throat, and then her breathing would stop, coming back only in fits. Sometimes the struggle would last a few minutes and sometimes a few hours. "I never knew when I would stop breathing, what would trigger it, or how long the attacks would last. The condition had such control of my life that I actually considered suicide."[6] Her husband lived in a state of high alert, never knowing when he would get a call from her coworkers. Whenever he did, he would take her home, where she would quickly climb into bed. "For the better part of two years, I slept," said Laurie. "It was the one thing I could do that would stop an attack."

Laurie was referred to the Cleveland Clinic, where Dr. Milstein strongly suspected vocal cord dysfunction. This meant that instead of moving outward with inspiration, her vocal cords moved inward, blocking airflow and creating shortness of breath. Using a very small video camera, he examined her vocal cords, watching them as she breathed and made sounds in different ways.

What set Dr. Milstein's approach apart from that of the other physicians who had examined Laurie's vocal cords was patience. Instead of just taking a quick look, he calmly sat there with his camera on top of her vocal cords. Finally, her vocal cords started moving irregularly, and at last Laurie had a diagnosis that could be worked with. Dr. Milstein prescribed breathing exercises, to be done twice a day. The exercises were simple, ones that could be learned in any yoga class: easy breath in through the nose five seconds, hold the breath two seconds, then easy breath out. Creating an awareness of the breath gave Laurie some control over it, and she could then maintain airflow during times when her vocal cords were moving inappropriately, buying time until they settled down.

Using these techniques, Laurie also taught herself the Buddhist and mindfulness principle of accepting an experience without judgment. If shortness of breath occurred, she reminded herself that it wasn't a problem with her lungs, nor was it an asthma attack; rather, it was merely an experience. Eventually the bouts of shortness of breath and panic ceased. Laurie insightfully calls her larynx "overprotective," the exact right word. We know that, based on its location, the larynx is under constant stress and must protect itself as we go about eating, breathing, coughing, and talking.

From an early age, Aubrey Murray was athletic. She began swimming at age eight on her local YMCA after-school team, then started to truly excel in high school, in Jensen Beach, Florida. Not surprisingly, Division I universities knocked on her door, offering her an athletic scholarship.

In college, Aubrey juggled a busy swim schedule with her chosen scholastic pursuit of nursing. Then, seemingly without warning, in the fall of her junior year she became acutely short of breath when swimming. She attempted to push through the difficulty, but things came to a head at the end-of-the-year championships while she was swimming her best event, the 200-meter freestyle. She began to slow down in the second half of the race, and by the time she got to the last lap she couldn't keep going. When her teammates and coach pulled Aubrey out of the water, they heard audible wheezing and saw that she was blue. Her parents came out of the stands to take care of her, and fortunately, over the next ten minutes she recovered.

Aubrey visited one doctor and then another, trying to learn what was going on with her breathing. I was one of the doctors she saw, and I was stumped. All her X-rays and blood work were essentially normal. We discussed the possibility that it was asthma, but the inhalers we gave her were ineffective. Finally, I received an e-mail from Aubrey's mother informing me that she was taking Aubrey to the Cleveland Clinic in Ohio to undergo a comprehensive medical assessment.

At the clinic, Aubrey's mother showed Dr. Milstein a video from the time at the pool when Aubrey had to be pulled out of the water. Dr. Milstein heard the distinctive whine that is typical of an upper airway issue. Then he studied Aubrey's voice box, where he found that while her vocal cords were fine, the muscles overlying her airway were not functioning properly. Whenever she took a deep and forceful breath, the muscles of her voice box collapsed over her vocal cords, impeding airflow. This occurred only when she generated extreme negative pressure with her breath, as happens when swimmers turn their heads to the side and suck in air as quickly as they can to minimize head rotation time.

Dr. Milstein had Aubrey duplicate the exact breathing pattern of a swimmer in his office, then diagnosed her with "supraglottic airway collapse." He first prescribed breathing exercises, but ultimately Aubrey needed to undergo a laser procedure to burn away some of the extra tissue above her vocal cords that was causing intermittent obstruction. After the treatment, she was back in the pool doing laps at the breakneck speeds that had won her a scholarship. In her senior year, Aubrey broke the pool record in the 200-meter freestyle.

The Cleveland Clinic today resembles Dr. Crile's clinic in theory and practice, albeit much expanded. Here, Dr. Milstein works alongside ear, nose, and throat physicians, surgeons, internists, and ancillary workers who are not traditionally on the staff of an academic division. This includes audiologists, who focus on hearing; speech pathologists; and even dentists. These specialists all work collaboratively, allowing for the efficient sharing of observations among the various arenas. Communication between institutes is also encouraged, and Dr. Milstein regularly works with lung doctors seeing asthma patients.

Patients referred to the Cleveland Clinic for shortness-of-breath issues are often evaluated by many types of specialists, who share their findings with one another immediately. The entire evaluation is often done in a day or two, while in a typical practice, the workup by different specialists may span weeks or months. This model not only delivers the best patient care, but is also our best bet to stay a step ahead of the artificial intelligence (AI) explosion that appears imminent. When Dr. Crile and his colleagues opened shop almost a hundred years ago, medical knowledge doubled every 150 years. This number fell to 50 years in 1950 and to 3.5 years in 2010. With eight hundred thousand new medical articles being published every year, by 2020 medical knowledge is estimated to double every seventy-three days.[7] Humans will be challenged in keeping up with this explosion of knowledge.

Big companies such as Apple, Google, and Amazon are betting on AI and are busy designing diagnostic and treatment algorithms from all their big data. So far, however, attempts to apply artificial intelligence in the field of medicine have been disappointing, and examples of failure abound. Elaborate protocols were developed to help diagnose patients with clots in their lungs. The protocols incorporated a myriad of data, including pulse rates, blood pressure, and laboratory values. The protocols seemed to work, but when physicians were asked to estimate clot risk using their own clinical acumen, their judgment was just as good, if not better.[8] The computer program for cancer doctors developed by IBM uses AI algorithms to come up with the best treatment regimen. A review in *STAT News* in 2018 demonstrated "multiple examples of unsafe and incorrect treatment recommendations."[9]

Certainly, with the advancement of technology, change is coming to medicine. With places like the Cleveland Clinic showing the way forward, the human mind and, just as importantly, the human touch will continue to be a vital part of healing and scientific advancement. There is a flexibility inherent to human thinking that allows us to generate new and unique ideas. This type of thinking goes to the very nature of consciousness, which computers will never be able to mimic.

Medicine must continually update its mission and methods to deliver the best care. Integrating computers and exploring the usefulness of artificial intelligence is only logical. The Cleveland Clinic, however, demonstrates the enduring power of the human mind. This is especially true at the intersection of the breath, voice, and swallowing. Dr. Milstein was able to focus on Laurie Danders's and Aubrey Murray's respective voice boxes after he was told by the other specialists their other organs were in good working order.

Laurie and Aubrey are also not alone; the voice box is an organ often ignored when considering breathing issues. In a cohort of 148 athletes evaluated due to shortness of breath, none were initially suspected of having vocal cord dysfunction. After receiving a second opinion, this jumped to 70 percent.[10] Nonetheless, with the advocacy of Dr. Milstein and the clinic's collaborative approach, patient care is improving. A 2004 study found that the average time from symptoms to correct diagnosis in patients with vocal cord issues was 4.5 years.[11] Data collected by

Dr. Milstein more recently indicates that the gap has come down to a little over 1.5 years, a significant improvement.

The voice, created by the flow of the breath, is vital to our existence. It helps us communicate with each other, persuade each other, and love each other. As the public and the medical community learn more about laryngeal issues, we will be better able to ensure that everyone has a healthy voice.

Chapter 13

The Miracle of Lung Transplant

The lungs are in constant communication with the atmosphere and bear the burden of this interaction by continually breathing in dust, mold, toxins, bacteria, and viruses. Along with the skin, the lungs are the most common portal through which infections invade the body. This relentless stress makes curing many lung diseases difficult, and it also made accomplishing the first successful lung transplant a monumental task.

Not surprisingly, the lungs were one of the last organs to be successfully transplanted. The first kidney transplant occurred at Brigham and Women's Hospital in Boston in 1964, and the patient lived for eight years. The first successful liver transplant was in 1967 at the University of Colorado, with the patient living for over a year. The first attempt at a lung transplant was in 1963 at the University of Mississippi, under the guidance of transplant surgeon James Hardy. Alas, the patient died just three weeks after surgery, and two decades passed before any progress was made.

Today, many issues with lung transplantation remain. The outcome statistics are abysmal compared to transplants of other organs. Kidney transplant recipients enjoy a five- and ten-year survival rate of approximately 80 and 60 percent, while the outcomes for both liver and heart transplants are both about 75 and 57 percent, respectively.[1,2,3] Comparatively, the five- and ten-year survival rates for a lung transplant are 55 and 34 percent, and recently no significant progress has been made toward a cure for the complications that cause death.[4]

The constant exposure to the environment that the lungs must endure also compounds the problems of infection and rejection in lung transplant recipients. The kidneys and liver are nicely protected in the abdominal cavity, and the heart sits deep in the chest behind the thick bone of the sternum. The lungs are on the front lines. Before a successful lung transplant could be accomplished, several obstacles needed to be overcome. In the same way that most advances in medicine occur, overcoming these obstacles would require hard work—but also some luck.

After the failure of the first lung transplant in 1963, forty-three more attempts were made up to 1983, though none of the patients survived any significant length of time. The failed experiences demonstrated that two problems needed to be resolved in order for lung transplants to succeed. And in a fortunate coincidence, the solution to one proved to be a big part of the solution to the other.[5]

Lung transplant patients required a better immunosuppression drug than prednisone, the standard steroid in use. Prednisone casts its net too wide and causes too many side effects; something more precise was needed. Many of the failed lung transplant patients were also having issues with their airway anastomosis (i.e., with the connection between the new, transplanted bronchus and what was left of the existing one). Lung transplants involve three major surgical connections. The first two—the pulmonary artery and pulmonary vein—are for blood supply to the lungs; the third is for the airway. In the experimental days of lung transplant, the first two connections were not a problem, but the airway connection was. Almost all of the early lung transplant patients suffered a breakdown where the surgeon had sewn the new bronchus to the piece of the old bronchus that is higher up.

In the late 1960s, developing a new immunosuppressant was on the to-do list of pharmaceutical companies throughout the world. At the time, no "rational design" governed the development process of new medications. Scientists did not go into laboratories and draw out the structure of new drugs they thought might work, then build them. It was not for a lack of trying this approach, but for lack of success with it. From the time Alexander Fleming first discovered penicillin in 1928 after noticing that a fungus left to grow in a messy laboratory could kill bacteria, drugs were found in nature—as they still are, for the most part.

Over the years, drug companies have had different ways of mining nature for pharmaceuticals. Sandoz, a Swiss pharmaceutical company, would give employees plastic bags to take on business trips and vacations to fill with the local soil. In 1969, one of these employees brought back a bag of dirt from Hardangervidda, Norway. Hans Peter Fry, a biologist at Sandoz, found the fungus *Tolypocladium inflatum* living in the dirt, and from that fungus isolated an antibiotic that they named cyclosporine. But Dr. Fry was disappointed in its strength as an antibacterial, and soon gave up on the drug.[6]

Fortunately for the history of transplant, Sandoz also had an interest in immunosuppressive drugs, mostly for use in cancer. The field of transplant was revolutionized when Dr. Fry passed his results on to a colleague, who screened cyclosporine for its ability to suppress lymphocytes. Lymphocytes were not just implicated in a variety of cancers; they were also the main cells responsible for all types of organ rejection.

While many compounds had been found to be toxic to lymphocytes, what was new and unique about the fungal-derived cyclosporine was that it was *not* toxic to bone marrow or other organs, as previously screened drugs had been. The drug didn't work for cancer, but in tests in liver transplant patients published in 1980, it was a clear home run for the transplant world. Sandoz had a miracle drug, and the lung transplant world had an answer it had been waiting for since 1963. Now a physician was needed who could solve the problem of how to effectively hook up the old airway to the new one.

Dr. Joel Cooper was not initially attracted to thoracic surgery, but an accident in the operating room fortuitously sent him in that direction. Like many, Dr. Cooper contracted hepatitis from blood he was exposed to while performing a surgery during his residency at Massachusetts General Hospital in Boston. The gloves used in the 1960s were inadequate, and the suture material was very heavy, making a cut through the glove and into the hand frighteningly common. Dr. Cooper was taken off surgery rotations to recuperate and placed behind the microscope in a pathology laboratory. There, given some time to think, away from the breakneck pace of the operating room, he got interested in the airway and in pulmonary medicine.

After completing his residency and an advanced fellowship, he moved to the University of Toronto to head its thoracic surgery program. Initially, he did surgeries for lung cancer and infections, but after helping on a failed lung transplant in 1978, he gave transplantation his undivided attention. Not waiting for the pathologist's report, he did the autopsy on the transplant patient himself, finding a broken-down airway connection. The others present thought this breakdown was from rejection, from an influx of inflammatory cells into the connection. But Dr. Cooper was struck by the appearance of the sutures used to connect the old airway to the new, gleaming back at him as if they had been put in an hour earlier rather than several days. He appropriately concluded the sutures hadn't *endotheliolized*—they hadn't been incorporated into the tissue, because the tissue had failed to mobilize the appropriate cells to do what they normally do with any type of sutures placed for a cut anywhere in the body. The unusual appearance of the sutures was a sign that it had been lack of wound healing, not rejection that had kept the tissues of the old and new bronchus from bonding together.[7]

Dr. Cooper began studying the lung transplant procedure in dogs, and he noticed that, with the typical immunosuppressive drugs, the connection between the old and the new bronchus would become tenuous a day or two after surgery. The connection site looked wafer thin, the sutures didn't incorporate, and eventually the bronchial connection would weaken to the point of separation. These observations convinced Dr. Cooper he had successfully mimicked the findings of a lung transplant in humans. He had his control.

Then Dr. Cooper did what great scientists and physicians often do: he borrowed a few ideas, this time from another area of surgery, and applied them to his problem. It was well known in medical literature that wounds healed poorly when steroids were given to the patient, so surgeons generally avoided them at all costs. Dr. Cooper followed the transplant literature closely and saw the success liver transplant doctors were having with the new drug cyclosporine. This led him to substitute cyclosporine for prednisone in the immediate postoperative period.

The next idea he borrowed came from the literature of abdominal surgery. Within our abdominal cavity is a seldom-mentioned thick sheet of tissue called the omentum. It is made mostly of fat, but also has some immune cells. The omentum hangs down from the stomach like an apron and provides a layer of protection for our abdominal organs. It can move, and it will cover an area of infection or inflammation in the belly as needed. From his time in residency doing abdominal surgeries, Dr. Cooper knew that when belly surgeons did not have complete faith in a surgical connection, they would take a piece of omentum and wrap it around the surgical site to reinforce it.

Dr. Cooper tried his new ideas on his dog model, substituting cyclosporine for steroids and wrapping a piece of omentum around the bronchial connection. When the procedure worked, he resolved to try this approach on a human patient, aware that subtle differences in species can make huge differences in outcomes.

His team at the University of Toronto included Dr. Ron Grossman, a medical lung transplant specialist, and operating room nurse Marva Gilkes. Here they practiced lung transplant together in the human autopsy lab, getting even the number of sutures needed to tie up all of the different connections down to an exact quantity. With his team, the right immunosuppressive drug, and the correct surgical technique in place, Dr. Cooper submitted a proposal to the University of Toronto's review board for human trials in lung transplant, and the document endures today as the guiding document for selecting appropriate candidates: patients who are sick, but not too sick, who are suffering from their disease but are not totally debilitated, should be transplanted.

The hospital approved the proposal for five lung transplants, with an agreement to stop there and analyze results. Given the tremendous pressure, Dr. Cooper understood that they couldn't take any chances, that every detail of the entire procedure needed to be analyzed and optimized. He also admits that he was not focused on justice—on offering a lung transplant to anybody who was very sick. He was focused on utility, on pragmatism, on picking just the right person without any medical problems other than lung disease to ensure a success.

The group found its first candidate in a fifty-eight-year-old hardware executive named Tom Hall. His breathing problem had started slowly, as a cough and some shortness of breath when walking, but progressed quickly. His physicians diagnosed idiopathic pulmonary fibrosis, the relentless lung disease with no cure and no treatments. The lungs are normally the consistency of a dense sponge. With idiopathic pulmonary fibrosis, over a few years, they harden, slowly suffocating the body. By 1983, Tom Hall had spent two years on oxygen, and intermittently needed a wheelchair to get around.

Like astronauts, all lung transplant patients are vigorously screened psychologically to see how they respond to stress, and whether they are prone to giving up. Dr. Grossman, the medical transplant expert, told Tom Hall that the surgery wasn't a sure thing, that he would be the forty-fifth attempt at a procedure that had failed forty-four times. Thoracic surgeons throughout the world had given up on the procedure, deeming it too risky. Tom thought about it for a moment, didn't flinch, and replied evenly that he would be grateful for the opportunity to be number forty-five. Hearing those words, the team knew they had their man.[8]

If there is a downside to the miracle of organ transplant, it is that somebody has to die, usually somebody young, for the procedure to be a success. The best organs are young organs. Tom Hall's new lungs came from a thirteen-year-old Quebec boy who was on life support but had been declared brain dead after a car accident.

In his determination to leave nothing to chance, Dr. Cooper decided before accepting Tom Hall for transplant that any donor would have to be physically located at the university hospital so as to minimize the organ ischemic time—the time the donor's lungs spent outside the body. It is well known that once organs are disconnected from their blood supply, they begin to die; death occurs within hours, possibly a day. The maximum acceptable ischemic time for lung donation wasn't known, but Dr. Cooper wanted to keep it as short as possible. For this next transplant effort, number forty-five, he demanded that the donor be in one operating room on life support, and the recipient in the operating room next door.

Dr. Cooper first heard of the potential donor because his kidneys were being offered. Dr. Cooper got on the phone to the hospital in Quebec and asked about the lungs; he was told the boy's father had also died in the accident, and the mother wanted to bury them together. Presumably that would not be possible if the boy's body had to go down to Toronto from Quebec. Dr. Cooper pledged to make it happen. His next call went out to the Canadian military, which agreed to airlift the boy down in their plane that could accommodate somebody on life support.

Arriving a few hours later, Tom Hall met Dr. Cooper in the preoperative area, and they spoke a final time about the operation, its risks, and benefits. Tom nodded again in approval. In the operating room, he was put into a deep twilight zone, and a tube was placed in his throat. The surgery started after midnight and progressed smoothly and quietly. The plan was to transplant just one lung, thinking that would be enough, and would shorten surgical time. They calmly removed the lung from the young boy in one room and carried it over to Tom Hall's open chest in the other. Dr. Cooper sewed up Tom's artery, his vein, and then his bronchus in a few hours. He wrapped his airway connection in omentum. They let the blood flow in, along with the air, and on gross inspection the lung appeared to function normally, with no air leak. Dr. Cooper stitched up the gaping hole in Tom's chest cavity, and they wheeled him up to the intensive care unit. They kept the breathing tube in, wanting to wait a few days to let some healing occur before they let Tom breath on his own.

With Tom successfully situated in the ICU, Dr. Cooper had to immediately deal with another emergent situation. The donor, the boy from Quebec, had been appropriately taken off life support after his lung had been extracted. The Canadian military, who had so generously brought the boy from Quebec, now said their regulations prevented transporting a cadaver. Dr. Cooper, aware of the promise he had made to the grieving mother and wife, frantically called around to local private airline companies. After a few pleas, he found one willing to make the trip. He paid the bill with his own money, and the boy was buried alongside his father.

For the first week after the operation, the main group of five doctors cancelled their other clinical duties and traded twelve-hour shifts, each one determined not to have something bad happen to Tom Hall on his watch. The tubes inserted in his body came out one by one, starting with the breathing tube on day four. Dr. Cooper took a camera and looked into his lungs, where he saw the bronchial connection holding steady. Tom Hall started physical therapy, and a month after the transplant his rehabilitation therapist told him it was time he pushed himself around in the wheelchair. Even more important, she told him he could do it without oxygen. Soon he went home, and returned to work just a month after leaving the hospital, a huge sign to the transplant team that they had achieved their goal. At the one-year anniversary, the doctors, Tom Hall, and his wife gathered at the hospital for a celebration with cake and coffee.

Joel Cooper and his team went on to build one of the most successful and innovative transplant centers in the world. Their procedure spread to other medical centers. In 1990, seventy-six lung transplants were performed in the United States. In 2018, there were 2,530.[9] The number of transplant centers today has grown to sixty-five in the United States. The impact of this procedure on tens of thousands of lives is unimaginable.

Tom Hall lived another seven years after his transplant, finally succumbing to kidney failure. He described those as his gravy years, like a whole new life.[10] Remarkably, at seven years, Tom still beats the expected survival of a lung transplant recipient today, which is about six and a half years. This is despite all the experience we have built up and the new and better medicines we have developed since.

Although average survival in lung transplant has slowly improved over the years, the numbers still don't look as good as those for liver or kidney transplants because of chronic rejection. Chronic rejection is different from acute rejection. Acute rejection occurs when the body decides it does not want this foreign tissue, and it sends massive numbers of white blood cells, typically lymphocytes, to attack the lung. A lung transplant patient will become acutely short of breath, even have a fever. A chest X-ray will show the white of inflammation within the lungs, where it should be black with only air. Fortunately, high-dose steroids help drive out all of the invading lymphocytes, and death from acute rejection is rare. After the storm passes, the clock appears to reset, and commonly the body then accepts the lungs.

Chronic rejection is caused by a completely different mechanism. It generally happens after the first year, whereas acute rejection occurs in the first few months. With chronic rejection, there is no huge influx of inflammatory lymphocytes, but just a slow scarring that occurs around the airway from the activation of fibroblasts, the main scar-tissue producing cells. As in IPF, steroids and other anti-inflammatory medications have no effect on the fibroblast. Pirfenidone, the drug approved for IPF, shows some promise in early studies, and larger ones are ongoing.[11] The underlying cause of chronic rejection is unknown; perhaps some toxic inhalant or typical mild virus is not being handled appropriately by the new lungs. It takes a lot of work to bring the *pneuma* into the body, and this comes at a price, especially if normal immune mechanisms are blocked by rejection medicines.

That said, even with the specter of chronic rejection in the background, today 92 percent of lung transplant recipients said in one survey they would do it again, and 76 percent were highly satisfied with the procedure.[12] If only we could find enough organs for those who need them. One solution to the shortage is to increase the percentage of lungs that can be harvested from potential donors. Lungs have by far the worst recovery rate from donors, and only 20 percent of lungs from donors are usable, much less than harvest rates for other organs.[13] They are a delicate organ and easily damaged. A new device, the ex-vivo lung perfusion machine, was developed at the University of Toronto to improve the quality of suboptimal lungs by perfusing them with high levels of oxygen and a nutrient-rich fluid before putting them in the recipient. Early results have been promising.

Further down the line, the science of stem cells and lung regeneration may eliminate the need for any type of foreign transplant. The dream is to take a single cell from a patient with failing lungs, bring it back to an undifferentiated state as Dr. Darrell Kotton has shown us is possible, and then generate all of the different types of cells that make up a complete lung. Growing the different cell types is a big challenge, but an even bigger one may be making the scaffolding to house all the different cells. The lung has an underlying acellular structure, termed the *extra-cellular matrix*, much like the frame of a skyscraper. As a vine attaches and grows on a lattice, this scaffold gives the cells and vasculature of the lung something to latch onto as they replicate and grow. Scientists are experimenting with two different approaches to build this scaffold: the first uses novel proteins to build it from scratch, and the second by stripping the cells out of an animal lung. If successful at total lung regeneration, it would eliminate any rejection issues and be an upgrade from a transplant of somebody else's lungs that already have years of wear.

PART IV

LIFE, LOVE,
AND THE LUNGS

Chapter 14

The Greatest Medical Story Never Told

C ystic fibrosis (CF) brings together the three main themes of this book—the central importance of the lungs, the courage of patients afflicted by a devastating illness, and the importance of hard work, intelligent observation, and collaboration in the advancement of medical science.

In the mid-twentieth century, people born with CF had an average life expectancy of less than five years, but that figure is now approaching fifty.[1] The story of this progress is not just a case study of how to begin to cure a disease. One day in the not-too-distant future, it will likely be a case study of how to end a disease. The achievements of the scientists in this story, and of the patients and their families, are nothing less than extraordinary. The unmasking of this disease is a remarkable tale.

August 25, 1989, was a very important day for CF patients. That day, Francis Collins, a tall, lean man with a mop of brown hair and an unruly mustache to match, stood in front of bright lights and twenty cameras at a much-publicized press conference. Already a successful scientist at the University of Michigan, and a future director of the National Institutes of Health, Collins had the nervous sweat of triumph dripping down his brow as he stated that the underlying genetic defect of cystic fibrosis had been discovered, headily declaring that a "base camp" had been set up on the Mount Everest that was cystic fibrosis. The summit of the mountain—a cure for CF—was perhaps a dozen or so months away.

But months turned into years, and years turned into decades, and the promise of a straightforward hike up the mountain did not materialize. The CF patients, their parents, and doctors around the country may have heard the press conference, but the human body had not taken notice. Although the genetic defect had been discovered, no big breakthrough in therapies had occurred, and progress had stalled.

A genetic disease, cystic fibrosis has likely been in existence for thousands of years. A description of what is almost certainly CF emerged in medieval times, when people were known to say, "Woe is the child who tastes salty from a kiss on the brow, for he is cursed, and soon must die." The first known autopsy on a CF patient was in 1595, by Peter Pauw at the University of Leiden in the Netherlands. He performed a postmortem on a "hexed" child and found the child's pancreas replaced with fat: "swollen hardened (and) gleaming white."[2]

Later, in the nineteenth century, more descriptions of what was likely CF were documented, as the medical use of autopsies became an accepted way to learn about disease. Dr. Carl von Rokitansky, at the University of Vienna, performed some thirty thousand autopsies from 1830 to 1878. In one case, he described an intestine blocked from impacted stool in a baby who had died just after birth. This pattern of intestinal blockage almost always indicates CF.

Over the ensuing years, other descriptions of infants who likely had CF emerged in the medical literature, most of them having died from malnutrition. What caused most of the mortality in CF then was malabsorption of food. To help us digest nutrients, our pancreas produces enzymes that break down the fat, protein, and carbohydrates of a meal so they can then be absorbed by our small intestine. Cystic fibrosis patients are usually born with a pancreas that is already destroyed, and without the replacement enzymes we prescribe today, these babies would have been unable to absorb nutrients, leading to a failure to gain weight and death within a few months.

The number of case reports listing the classic signs and symptoms of CF picked up in the 1930s. The people who wrote these papers were almost certainly describing CF, but they didn't realize they were observing a distinct entity, and CF was often confused with celiac disease, another sickness that causes malabsorption of food. One of these reports came from Dr. Guido Fanconi, a Swiss pediatrician who put his stamp on many areas of medicine. In 1936, he described two children with both lung mucus and pancreatic failure. One died at ten months and the other at three years. The term *cystic fibrosis* was first used in this report, referring not to anything in the lungs but to the pathological state of the pancreas—filled with fatty holes (cysts) with fibrosis mixed in.

Dr. Fanconi thought he was naming something new, but he didn't get credit for it. As is so common in the scientific world, scientists get credit for a discovery only by convincing the rest of the world their findings are original. The person who did that, and whose 1938 paper describing cystic fibrosis stands as the beginning of the academic study of CF, was Dr. Dorothy Andersen of Columbia University.

Born in 1901 in Asheville, North Carolina, the only child of a Danish father who died when she was thirteen, and an American mother with chronic health issues, Dorothy Andersen was tasked with much of the work required to keep the family going. Her mother passed away in 1920, and she then was able to dedicate herself to medicine, putting herself first through Mount Holyoke College and then through Johns Hopkins School of Medicine, completing her MD degree in 1928. She went on to a surgical internship at Strong Memorial Hospital, in Rochester, New York, but then was denied a residency because of her gender; undeterred by the discrimination, she forged ahead in the discipline of pathology, taking on the position of research assistant at Columbia University in 1930, and a few years later joining the faculty.[3]

In 1935, during an autopsy of a three-year-old with presumed celiac disease, Dr. Andersen first suspected something unusual. She found the pancreas, ordinarily normal in this condition, in a diseased state. Much like what was described by Pauw in 1595, this pancreas was a gleaming white. Under the microscope, its normal structure was mostly replaced with fatty tissue, along with elements of cysts and fibrosis—the organ was clearly not making the appropriate enzymes for digestion.

Her scientific senses alerted, Dr. Andersen sought out other children who had died with digestive issues and an abnormal pancreas. She found some at her own institution and some that she learned about from other pathologists throughout the country. In the end, she had forty-nine cases, which she wrote up in 1938 and published in a report entitled "Cystic Fibrosis of the Pancreas and its Relation to Celiac Disease."[4]

The paper caused a huge stir in the medical community throughout the world, clearly making the case that an unrecognized disease was occurring in infants and toddlers. That Dr. Andersen had included the work of physicians from other hospitals in the report surely interested those other institutions, but what likely ensured the paper's impact was the number of cases Dr. Andersen reported on, which was exponentially larger than anything published before. Because of her paper, physicians across the globe began realizing the syndrome Dr. Andersen was describing had nothing to do with celiac disease, and that a disease, undiagnosed and unrecognized, was right under their stethoscopes.

Over the next two decades, Dr. Andersen continued her clinical work on CF, establishing herself as the world's leading expert. In 1946, with her colleague Russell di Sant'Agnese, she published the first account of the use of an antibiotic to combat CF respiratory infections, and that year she also correctly designated CF as an autosomal recessive genetic disease. This meant that if one parent had the gene with the CF defect, a child would be fine, but if both parents passed on the defective gene, the child would be afflicted. (This is in contrast to an "autosomal dominant" disease, in which a defective gene from one parent is enough to cause disease.)

With these advances, Dr. Andersen and Dr. di Sant'Agnese helped turn cystic fibrosis from a postmortem description into something that could be diagnosed in the living. They developed the first test for CF, which checked the small intestine for the absence of a particular enzyme, and then developed another test, discovered mostly by chance, and brought about by a brutal heat wave in New York City in the summer of 1948.

The city then did not enjoy widespread air conditioning, and young children were being admitted to Columbia University Hospital because

of dehydration. Dr. di Sant'Agnese noticed that many of them had a diagnosis of cystic fibrosis. He hypothesized that if these young CF patients were more prone to excreting salt in their sweat, water would follow this salt, which would lead to the dehydration they were seeing. He collected their sweat and tested it, and the levels of salt were off the charts. (Now we know where that medieval saying, "Woe is the child who tastes salty," comes from.) Today, even with advanced genetic tests, testing the sweat for high salt levels is still the gold standard when making a diagnosis.

The science and treatment options for cystic fibrosis advanced rapidly in the decades after Dr. Andersen published her 1938 study describing the disease. The first big development was the use of pancreatic enzymes to do the job that the fibrotic and diseased pancreas could not. Within each capsule are granules of lipase, amylase, and protease, the three enzymes we need to breakdown fat, carbohydrates, and protein into absorbable units. This immediately changed the outcomes, as CF babies could finally process nutrition with the help of these enzymes to break down food.

With the pancreatic-enzyme deficiency much improved, babies with CF were living longer, but now they began encountering issues in the lungs, with excessive mucus production and subsequent colonization by aggressive bacteria. To combat these bacteria, CF physicians and scientists began experimenting with new antibiotics, such as erythromycin, terramycin, and aureomycin, using them as Dr. Andersen had first used them, both inhaled and intravenously. These drugs provided huge benefit to the lungs, keeping the disease at bay for a time so the patient's lungs could function and not buckle under a heavy load of bacteria, mucus, and pus.

With these new medicines, the life expectancy of CF patients increased from six months of age in 1938, to about two years in 1950, to the landmark age of ten in 1962. As important as the science was, though, something else happened at this time that would affect the course of CF more than any one researcher. The one group of people

even more passionate about finding a cure for CF, even more emotionally invested than any researcher or physician or social worker, were
the parents of the patients. In Philadelphia in 1955 a gathering of them
formed a national advocacy group, which would change not only how
we treat CF, but how we think about, treat, and solve diseases in general.

Among this group, two of the first and most invested members
were Milton Graub and his wife, Evelyn. Milton earned his medical
degree from Hahnemann Medical College in Philadelphia in 1945 and
started a practice in pediatrics nearby. Their son, Lee, was born in 1948,
but he did not seem healthy; difficulty with weight gain appeared first,
and then lung infections. Milton had recently heard about the newly
described disease, cystic fibrosis, and took Lee, two years old then, to
Dr. Andersen, who confirmed the diagnosis. By this time, Evelyn was
pregnant with their daughter, Kathy. She too would be diagnosed with
CF, this time shortly after birth.

The shock of the diagnoses hit the Graubs very hard, as Milton
remembered: "We were completely devastated that our children had a
disease that not only was incurable, but that very few people had ever
even heard of."[5] At the time of Lee's diagnosis, pancreatic enzymes and
a few antibiotics were the only treatment options. Knowledge about
airway clearance was just beginning to emerge.

The first thing the Graubs did in response to their lives turning
upside down was to contact other parents in the Philadelphia area whose
children had cystic fibrosis. They correctly assumed there would be
power in their numbers and in a unified voice. Together they formed a
local advocacy group in 1952 to create awareness about cystic fibrosis. In
the early days, Milton Graub would visit every family who had received
a diagnosis of CF within two hundred miles of his house, making sure
their questions were answered and that they had access to the medicines
and therapies available. Other parents began their own visits, in a classic
grass roots effort.

The local advocacy group grew and, in 1954, had its first fund-
raiser and guest speaker. In 1955, with the help of the Graubs and
families from many other cities, a national charter called the Cystic

Fibrosis Foundation was established. Local chapters were established in cities such as Cleveland, Baltimore, and Boston. The group began accumulating funds and directing money to projects and ideas to move research forward. Because of targeted fundraising efforts in multiple cities, and the heartrending lethality of the disease, millions of dollars poured in, as they would for decades to come. The foundation held its first scientific meeting in 1955 in Iowa, with many of the top scientists of the day attending.

Beyond fundraising and allocating grant money, the CF Foundation did something even more important, something way ahead of its time. In 2013, the Oxford English Dictionary added the phrase *big data*. The term means slightly different things in different fields. In the tech world, for instance, big data refers to using vast amounts of consumer information to predict what a single individual may like. In medicine, big data can refer to the collection of clinical information on a large number of patients, which can then be used to help build models of outcomes.

Way before tech companies were using big data to predict what ads would draw a response, the CF Foundation was collecting data on every CF patient in the country. From the height and weight of patients, to the medications taken, to the age at diagnosis, to the specific bacteria harbored in the lungs, the data was exhaustive—and collected multiple times a year. This information would be key to elucidating trends in patients, enabling the development of guidelines on what medicines were needed and at what age. The data would also play an important role in allowing researchers to study new medicines when they became available, elements essential to the process of finding a cure for a disease. Doctors in other specialties throughout medicine have seen the wisdom in this big-data approach, and in recent years many others have begun to track their patients based on the model the Cystic Fibrosis Foundation pioneered decades ago.

The CF Foundation also helped establish specialized cystic fibrosis centers, where CF patients could get the best and most cutting-edge treatment from physicians with specialized training and understanding of the disease. These clinics of excellence were always set up at academic medical centers, with the first two established in 1961, at Columbia, under the guidance of Dr. Andersen, and in Boston, under Dr. Harry Shwachman. The number of centers grew rapidly; one year later, thirty more were established, and today there are more than one hundred and thirty.

What made these specialized centers even more attractive to families was that patients would be seen not only by a physician with specialty training in CF, but also by a nutritionist, a respiratory therapist, a social worker, and a research coordinator. Cystic fibrosis is a multisystem disease that deserves a multisystem approach. This model for care was unique at the time, and it remained so for many years, until other specialties began appreciating its advantages. Now, subspecialty clinics for COPD and interstitial lung diseases and cancer are emulating this model. As evidenced by the outcomes, and the persistent uptick in life expectancy for CF patients, it is a model that works.

The Graubs' support for the CF community continued even after their son, Lee, died at the age of ten in 1958, followed by their daughter, Kathy, at the age of eighteen in 1969. Milton and Evelyn's prodigious fundraising also helped establish the Kathy and Lee Graub Cystic Fibrosis Center at the Schneider Children's Hospital in Tel Aviv, Israel, in 1995, which has delivered care to hundreds of children over the years.

The names of Dr. Graub, his wife, Evelyn, and all the parents, friends, and advocates they worked with in those early years won't be found in any scientific textbooks. But by helping to establish the CF Foundation, they are the ones who truly changed the course of the disease, through advocacy and fundraising, and by having a well-defined mission that they held to tightly.

In the late 1970s, the National Institutes of Health offered to fund research grants for projects related to cystic fibrosis, and with five years of support, each grant represented a significant amount of money. Thirty-five applications came in for the prestigious award, and the NIH sat down with a study group of outside scientists to decide who should get the money. After reading the grant proposals, they reached a unanimous decision that none of the projects was worth funding.

One of the people at the NIH who judged the science as sub-par was Bob Beall, who came to the NIH in 1974. He initially had no intention of getting involved with an orphan disease (one that affects fewer than two hundred thousand people) like cystic fibrosis. As so often happens, a human story changed his mind.

On a cold February day in Bethesda, Maryland, in 1977, Beall's boss at the NIH asked him to help manage the grants and NIH pro-gram for CF, and to attend a meeting in La Jolla, California, of leaders from the CF Foundation. Beall couldn't spell cystic fibrosis at the time, but Southern California in February sounded good. The weather was predictably nice, but the science he saw when he spoke with leaders from the foundation was, as he had previously determined, not up to par. What deeply moved him, though, was when he spoke with family members during breaks and saw their desperation for answers.

Their strong mission and sense of purpose convinced Beall to get involved with CF. Initially, he worked on CF within the NIH, and then, in 1980, he joined the CF Foundation full time to lead its research efforts. A few years later, he was elevated to president and CEO, posi-tions he would hold for twenty-one years. Under Beall's guidance, the foundation took the next big leap forward toward curing CF and became the envy of other foundations in medicine.

Beall worked to bolster the science supported by the Foundation, specifically setting up projects to investigate what was happening at the cellular level that caused destructive mucus to build up in a CF patient's lungs. Doctors at clinical sites across the country wanted money to take care of patients, but Beall convinced the board members that they needed to figure out the basic cellular defect, and that investing heavily in science was the way forward. The board was hesitant, since a scientific research project like this, on the scale Beall was proposing, had never been funded by a nonprofit foundation. The government and universities had funded major research projects, but for a relatively small foundation, it was a huge financial risk. Funding individual grants was one thing, but paying for large salaries and major infrastructure at a university was going to be expensive—in the tens of millions of dollars, eating up a significant portion of the foundation's resources. But from Beall's perspective, if they were going to get to their goal, which was a cure, then they would have to take some big risks. The board of the CF Foundation finally agreed.

In 1982, the first three research centers at major universities opened, at the University of North Carolina, the University of Alabama, and the University of California, San Francisco. Named Research Development Program (RDP) centers, their mission was to work together to fill in the gaps in knowledge about cystic fibrosis. From a scientific perspective, they all agreed they needed to "move upstream."

In science, *upstream* and *downstream* refer to where things are happening at the cellular level. Farthest upstream are the genes, made up of DNA, which contain the physical code for how a protein is constructed. Moving downstream, RNA is made directly from DNA. Further along, proteins are made from RNA. For years, all of CF care was focused downstream, on clearing out mucus and treating infections with antibiotics. But working downstream doesn't cure a disease, doesn't fix the defective DNA, the RNA, or the protein. That needs to happen within the cell itself, way upstream.

The RDP centers were a success, with the number of scientific papers doubling, and then tripling, in the following years. But what Bob Beall really wanted was to find the genetic defect. Cystic fibrosis was clearly a genetic disease, since it was known to run in families and be passed from one generation to the next. He understood that if they were going to find a cure, they needed to locate where the defect on the DNA was happening and identify the defective protein this DNA was encoding. Presumably, this protein was involved in mucus production and also in the functioning of both the lung and the pancreas. The gene and its concomitant protein were at the crux of the entire disease.

The first breakthrough in the search for the gene came in 1985, when Lap-Chee Tsui, at the University of Toronto, isolated the defective CF gene to chromosome 7.[6] In the nucleus of every cell of our body, there are twenty-three pairs of chromosomes. Depending on the location of the cell, only certain parts of the DNA are activated, and these parts produce the proteins needed for that cell to function in that specific organ. Proteins then do all the work the cell needs to stay alive—breaking down carbohydrates, regulating salt and water, and producing energy, among many other duties.

Our DNA makes proteins in a simple manner, first outlined by Francis Crick and James Watson in 1954. All our DNA is made up of one of four base pairs, termed A, T, G, and C. These base pairs are grouped in sets of three, and string out together in various patterns. Based on their arrangement, certain amino acids will be pulled together to form a protein. For example, if a particular part of the DNA has the three base pairs GCC, the amino acid alanine will be pulled. If the next three base pairs are GAC, then aspartic acid will join the growing protein molecule, next to alanine. Thinking backward, all proteins are made up of a string of amino acids, and the type of amino acids pulled for a specific protein depends on the configuration of base pairs in DNA.

Worth noting, however, is that only one base pair separates the code for the two amino acids mentioned above. If the DNA reads a C where an A should be in a specific spot, then alanine will be added to the protein instead of aspartic acid. A single wrong amino acid from a mistake in a single base pair in the DNA (termed a *point mutation*) can produce a malfunctioning protein—even though our proteins consist of hundreds, and often even thousands, of amino acids.

In cystic fibrosis, it was believed that some very specific part of our DNA, somewhere in one of our twenty-three chromosomes, was broken. At a cell-biology level, this meant that somewhere the DNA was constructed of the wrong base pairs, that the specific arrangement of A, T, G, and C was off. Since base pairs encode amino acids, the wrong amino acid would be added to the protein being built, leading to aberrant function. Presumably, in CF this protein was involved in the regulation of mucus production in the cells of the lung and pancreas.

The human genome is a big place, and our twenty-three chromosomes contain approximately three billion base pairs, which encode the twenty-one thousand proteins we use to stay alive. This is a vast area in which to look for a defect that may be only one base pair off. To narrow things down, Lap-Chee Tsui's lab analyzed the DNA in families in which CF was prevalent and looked for broad similarities in their genetic material. Dr. Tsui found that all the families seemed to have similar changes happening in a specific part of chromosome 7. The DNA that exhibited the changes was of no consequence (only 1 percent of our DNA makes proteins; the job of the rest is still not clear), but it was a hint that something was different in this area in the families in which CF was common.

With localization down to chromosome 7, Dr. Tsui had whittled the amount of DNA to analyze from 3 billion base pairs down to 159 million, still a daunting challenge. To get granular detail would not be an easy task, and some thought it would be impossible with the techniques available. Scientist Francis Collins at the University of Michigan, excited about the work done in Dr. Tsui's lab, believed some methods he had used, known as *chromosome walking* and *chromosome jumping*, might be helpful. He proposed to Dr. Tsui that they not only collaborate but join labs for this project. In a remarkable display of solidarity, for several years the two labs shared data and methods in their singular quest to discover the CF gene. The five-hour drive between Ann Arbor and Toronto became familiar to both scientists. As an incentive for their quest, they bought a bottle of Canadian whiskey and put it on the shelf, with the promise to open it only when the gene had been found.

Dr. Collins motivated the people in his lab in unique ways. Working in Michigan, where barns were plentiful, he easily found out that the weight of an average haystack was eight tons. An average needle, he learned, weighed two thousand milligrams. Proportionally, their search for the defective base pair among the millions, they were quite literally looking for a needle in a haystack. Next, Dr. Collins took a picture of himself in a barn, on top of a pile of hay, a chicken and a pitchfork to his left and old wooden rafters overhead. In his hand he held a shiny silver needle. The message was clear: this was doable.

Chromosome walking and chromosome jumping were methods used to analyze DNA and identify the exact base pairs in a specific area. With the technology available at the time, large pieces of DNA couldn't be analyzed: it was simply not feasible. Very small pieces could be analyzed, but they had to be cut up for analysis, and the process of cutting up the DNA made it impossible to then say where exactly the analyzed piece fit into the string.

To perform an analysis, scientists would start at a specific area of the DNA that they thought could be useful. For their analysis of chromosome 7, Collins and Tsui started at the area in which Tsui had located similar mutations in the noncoding DNA of CF families. They knew this wasn't the gene they were after, but they believed that gene was close by, that the two areas were linked. Tsui had been able to find this noncoding area of similarity because it was exponentially larger than the parts of the chromosome that do code proteins. But it was also just chromosomal nonsense, strings of repeating base pairs that happened to be different in CF families.

Still, it was a place to start. And from this starting point, Collins and Tsui began "walking" down the DNA, cutting up pieces of it at different intervals, and then analyzing the base pairs in each segment. Because they cut the same piece of large DNA at different places each time, there were overlapping sequences in the analyzed code, which they could use to string the entire chain back together in the correct order. As an example, one piece of DNA may have the sequence ACTCAG at the end of an analyzed segment, while another segment might have that same ACTCAG sequence at the beginning. These two pieces could then be fit together to get the sequence of a larger piece of DNA.

After "walking" in one area, they would then "jump" to another part of the chromosome and do the same thing there, cutting in different areas, analyzing the pieces for their base pairs, and then again fitting the entire puzzle back together by matching overlapping sequences, this time also fitting together the part they had jumped from previously.

Using this method of walking and jumping, Collins and Tsui slowly reconstructed the genetic material on chromosome 7, specifically around the area Tsui had identified previously. They used this technique not just on the genetic material of those with CF, but also, as a comparison, on the genes of people from families without CF, searching for any differences that might be significant.

The needle in the haystack was hard to find. Some genetic diseases have thousands of base pairs missing in the chromosome (large deletions), and naturally it is easier to find an error if thousands of base pairs are missing than if only one pair is missing. The malfunctioning protein of CF would not be caused by a large deletion. But slowly, through this painstaking process of cutting, analyzing, and fitting together, the puzzle began to take shape.

The answer came to Collins and Tsui while they were attending a scientific conference in May 1989 at Yale University, in New Haven, Connecticut. Every evening, the two scientists would go back to their dorm and review the faxes of base-pair analysis from the day's work in the lab. One rainy night, they picked up a rolled-up piece of fax paper from the floor and found their answer. A comparison of base pairs in a very specific area of chromosome 7 showed three base pairs missing in those affected with CF, a C, a T, and another T. In those without CF, these three base pairs were always there. The three missing base pairs were from a chromosome that would prove to be 188,702 base pairs long. From that chromosome, one single amino acid out of 1,480 gets left out of the protein, the protein doesn't fold or function properly to manage water and electrolytes, and the result is cystic fibrosis. Quite clearly, they had their defect, and their gene.[7]

Afterward, to fulfill a promise to themselves, they headed back to Toronto and opened up the bottle of Canadian whiskey to celebrate. The lab technician they asked to photograph their historic toast thought it strange that senior scientists were drinking hard liquor at ten o'clock in the morning, but the smiles on their faces, with stacks of medical journals in the background, showed pure joy. The hope in the CF community exploded with this momentous discovery. One eight-year-old CF patient wrote a diary entry the day the discovery was announced;

forwarded to Dr. Collins, it read, "Today is the most best day ever in my life. They found a jean for cistik fibrosis."[8]

The discovery of the gene responsible for cystic fibrosis connected with work done in the early 1980s on what was happening at the cellular level that was causing the problems with mucus in the lungs. Dr. Paul Quinton was one of the scientists doing this work. Paul had suffered from a cough and lung infections throughout his childhood in southeast Texas in the 1950s. Getting no answer from the physicians he saw, he did his own research and came upon the possible diagnosis of cystic fibrosis. He presented his hypothesis to a lung doctor in Houston, who saw the wisdom in this nineteen-year-old's insights and conducted a sweat test, which confirmed the diagnosis of CF. When Paul asked him how long he was going to live, the lung doctor told him honestly that he had no idea, that technically he should be dead already.

Paul took the doctor's honesty to heart and forged forward with research, first obtaining a PhD at Rice University, and then doing post-doctoral work at the University of California, Los Angeles. Here, his interest in physiology merged with a desire to figure out the cellular defect of the very disease that afflicted him.

When Dr. Quinton began his scientific work in the 1970s, it was known that something aberrant was happening in the cells of CF patients. The prevailing idea at the time was that some factor in the blood of CF patients caused the electrolyte abnormalities and fluid imbalances seen in the affected organs, such as the lung, pancreas, and skin. Since salty skin was a clear sign of CF, Dr. Quinton thought this was a natural place to start, so he began investigating how both normal cells and cells in patients with CF move sodium and chloride, the two substances that make up salt.

He started looking at cells within the tissue of CF lungs, but they were too scarred to be studied effectively. He decided to focus on the sweat glands, which functioned abnormally but were free from scar tissue and not pathologically affected. For controls, he used the skin discarded from hair transplants of middle-aged men. For CF sweat glands, he got skin from local CF patients, but also from himself—the scars on his arms today are evidence of where chunks of tissue were removed. The experiments he conducted culminated in a 1983 *Science* paper that showed cells in the sweat glands of CF patients did not move chloride out appropriately. The paper also showed that this dysfunction was not caused by a defect in the blood, but rather by a defect in the permeability of the chloride channel of the cell.[9]

The research of Tsui, Collins, Quinton, and others eventually connected, and the story that emerged completed the chain of causality from gene to protein to disease. The loss of three base pairs in the genetic code leads to a single amino acid missing from the protein, causing the protein to fold defectively. The protein, called the cystic fibrosis transmembrane conductance regulator (CFTR), would normally move to the top of the cell and allow chloride to move out. When this system is broken, chloride gets stuck inside the cell, and the negatively charged chloride attracts the positively charged sodium. Since sodium and chloride combine to form salt, water is next attracted into the cell. Without the lubricant from salt and water on the top of the cell, mucus dries up and hardens, and bacteria, inflammation, and disease follow.

By the mid-1990s, the biochemistry part of the CF story was well understood, and a potential way forward also seemed clear. A new technology, called *viral vector transfer*, held the promise of a quick cure through gene therapy. Some viruses work by inserting their genetic material (either in the DNA or RNA form) into the DNA of a host, and then using the cell machinery of the host to transcribe their DNA into the proteins they need to create copies of themselves. They cannot replicate on their own, but they do have millions of years of experience exploiting the cells of other creatures, including humans. In fact, about 8 percent of our DNA is from viral DNA permanently left behind.[10] For CF, researchers proposed loading a normal copy of the CF gene into a virus's DNA, then letting the virus do the hard work of putting this functioning gene into a patient's DNA.

The problem was that although viruses have indeed been inserting their genetic material into ours for a long time, and are very successful at doing this under certain circumstances, our immune system has also been fighting back against this intrusion for an equally long time. For gene therapy in CF, loading the normal CF gene onto a virus was not the problem. The problem was figuring out how to evade the body's immune defenses and get the virus to deliver the normal copy of the CF gene to the cells of the lung. The exact circumstances needed for gene transfer to occur were elusive—and remain elusive to this day.

The failure of gene therapy in the 1990s was a huge blow to the CF community and the CF Foundation. Word had spread after the discovery of the gene in 1989 that a cure was not far off. But, as they so often had, the CF Foundation picked themselves back up and thought about the next big thing they were going to do, the next big risk they were going to take. Their resolve would take them to a place no disease had gone yet.

By the mid-1990s, a great deal of progress had occurred in CF treatment. New therapies were coming out, such as Pulmozyme and inhaled tobramycin, which helped suppress and clear mucus. Life expectancy was creeping up every year. But the CF Foundation realized, as did patients and their families, that all the therapies and improvement were just nibbling at the edges of the disease. Treatments were still focused downstream, and everybody knew the golden key to unlocking CF was to improve chloride movement from inside the cell to the top of the cell so the mucus didn't build up and the infections didn't develop in the first place.

At forty-three words, the mission statement of the CF Foundation is short, but the first twelve words are what a lot of people, including the patients, are the most interested in: "The mission of the Cystic Fibrosis Foundation is to cure cystic fibrosis." So, in the mid-1990s, with the hopes of gene therapy fading, the CF Foundation decided it needed to focus on the cellular middle ground of the defect. The foundation had worked downstream for thirty years, focusing on clearing the mucus out and treating infections. Then, for ten years, it had worked upstream, first identifying the defective gene and then trying to modify it back to its natural state. Now the foundation moved to the middle, focusing on the malfunctioning protein.

Even with the defective protein, the cells of a person with cystic fibrosis are still able to transport about one percent of the normal amount of chloride, sodium, and water out of the affected cells. Scientists believed that if the defective protein could be boosted to work at a 20 percent level or, even better, a 50 percent level, it would make a significant difference in people's lives, preserving enough lung function to enable someone with CF to live a long and healthy life. To get the defective protein to work somewhat better, scientists thought they could rely on protein modulators—pills that would manipulate the cell's apparatus to improve the quality and amount of defective CF protein that was being produced within the cell.

However, the problem was the protein modulators the CF Foundation was advocating didn't exist. In fact, there were no FDA-approved protein modulators for any disease. In the world of biochemistry, protein modulators belong to the domain of small-molecule science, the study of molecules that are small enough to both enter the GI tract and make their way into a cell to effect change. In the mid-1990s, small-molecule science was focused on turning things off, like activating a kill switch to stop replication of a cancer cell. There was no precedent to develop a molecule that turned something on, like what would be needed to augment production of the CF protein.

The timing, however, was auspicious. In 1990, the human genome project had begun to catalog the twenty-one thousand genes in our DNA that encode all the proteins we need to survive. This provided scientists with a lot of potential targets for manipulation. At the same time, advances in chemistry allowed large libraries of medicines to be tested very quickly. Some motivated individuals were needed to combine the new knowledge in chemistry and biology into something useful. Those individuals proved to be in a company in Southern California.

Aurora Biosciences was started in 1995 in San Diego. It caught the eye of the CF Foundation because of its unique ability to screen thousands of drugs at a time. The company had developed a technology, called *high-throughput screening*, that could test thousands of small-molecule compounds a day for effectiveness, while university labs were stuck screening a few per week.

The Aurora scientist who took the CF Foundation's call was Paul Negulescu. His administrator told him that Bob Beall from the CF Foundation was on the line for him. When he told his administrator to take a message, she didn't budge: "I don't think this guy is going to get off the phone until you come and talk to him."[11]

Bob Beall got right to the point. "I just saw a talk on high-throughput screening, and I want you guys to do it for cystic fibrosis. Can you do it?" Introspective by nature and painfully aware of the many pitfalls and blind alleys that science can go down, Negulescu responded with a measured "We can try," while remaining unconvinced of the true possibility of success.

With the support of the CF Foundation, Negulescu and his team began working on the problem. Their plan was to screen thousands of medicines by seeing whether they improved chloride conductance in CF cells in test tubes, then take whatever drugs they had found effective in the screening process to human trials. To give an idea of the predicted success rate of this approach, the Aurora team thought that more than a million compounds would have to be screened to get two or three small molecules worth taking into patient trials.

The project began in earnest in 1998, and with a $46 million investment by the CF Foundation in 2000, progress began to be made. First, cells with the same defect seen in CF patients needed to be developed in order to test these small molecules outside the body, a crucial first step. A human lung epithelial cell from a CF patient would be the best cell to test the drugs on, but obtaining the needed number of cells was not feasible; growing human tissue with the CF defect was too onerous to be undertaken on the scale needed. Negulescu and his team examined twenty existing cell lines of epithelial cells from different animals, ones being used in other labs and known to be amenable to research needs. For a cell line to work, the team would need to be able to change its DNA to include the CF defect, and then make sure the cells stayed alive while researchers inserted drugs into them to test for effectiveness. This took months, but it was far faster than having to develop their own cells from scratch. They got lucky: one out of the twenty cell lines worked, the epithelial cells from the thyroid of a rat. After outfitting these cells with the genetic defect of CF, the scientists could inject a small molecule into the rat thyroid cell and see whether it improved chloride conductance.

Then the actual physical machinery for doing the experiments had to be built, since screening a million compounds by hand was not realistic. Each day, thousands of tubes had to have microscopic amounts of solution and drug inserted into them at variable times. It would not be feasible for a person, or even group of people, to do this work on the scale needed. Laboratory robots had not yet been created, so the team purchased robots from the automobile industry. Built for bending steel, the machines were a little rough, and many test tubes were crushed and cell plates dropped. Eventually, the engineers got it right, and high-throughput screenings for drugs that could fix the CF protein were up and running.

Screening these thousands of compounds turned into a massive project, and Paul Negulescu acknowledges the only way he and his team got through it was with the enormous support they received from the CF community and foundation. One early interaction Negulescu had left a huge impression: the visit of a four-year-old CF patient to the laboratory the day before a fundraising walk they would all be doing together. The boy told him, "I just want to thank you guys for working on my vitamins." The mixture of hopefulness, vulnerability, innocence, and gratitude displayed by this child with a devastating disease stayed in his heart, and a combined deposit of willpower and obligation helped him persevere through whatever blind alleys the science would lead him down, whatever frustrating days and nights of letdown and setback he would encounter.

Out of more than one million compounds screened, fewer than one hundred potential small molecules worked to improve chloride conductance in their rat cells. Having identified its candidates, the team moved to test these drugs on actual human bronchial cells that had the CF defect, a process that narrowed the possibilities down to a handful. Two in particular stood out, and the medicinal chemists went to work on improving potency (how many grams of a drug need to be taken) while maintaining efficacy (how well a drug works) as well as absorption ability. They had to reject the most potent compound because it lacked other needed properties, but in the end they had the first drug they thought could be brought into clinical trials. Called VX-770, it would be only for those CF patients with the G551D mutation. This was not the genetic change Tsui and Collins had found in their lab in 1989. Later discoveries showed that the CF protein, with its 1,480 amino acids and 180,000 base pairs, could have mutations at hundreds of different places to cause the malfunction. G551D was present in only about 4 percent of the total CF population, but it would be a huge breakthrough for the field to have anything at all—to have discovered a small molecule that would fix something that was broken.

The story almost ended there, because in 2001 the biotech company Vertex Pharmaceuticals, based in Cambridge, Massachusetts, acquired Aurora. Like Aurora, Vertex focused on small-molecule drugs. But what had caught Vertex's eye was not Aurora's CF research but its high-throughput screening technology, which Vertex hoped to use to develop blockbuster drugs for cancer and hepatitis, which offered a market of millions of patients. When the new management reviewed Aurora's contract with the CF Foundation they discovered that the world of CF was merely thirty thousand patients in the United States and seventy thousand patients worldwide, and they seriously considered cancelling the agreement.

Vertex was also concerned about the idea of working with a nonprofit organization like the CF Foundation. Such a large investment from a private nonprofit in a for-profit drug company, known nowadays as venture philanthropy, was nonexistent at the time, a model without precedent. Bob Beall remembered when he began to sell this idea to pharmaceutical companies, the conversations were short. "First, your disease is too small. Second, you're a charity," they would tell him. Also relevant, in the late 1990s, bringing a drug to market cost about $800 million and took about fifteen years. Companies needed to think they could make their money back. Bob felt lucky with the relationship with Aurora. Now that was in jeopardy.[12]

One thing that helped Vertex make its decision to continue was the enthusiasm of its lead scientists, most of whom had come from Aurora. At a meeting with the chief scientific officer at Vertex, scientist Fred Van Goor went through presentations on various drugs, giving them each a sell. Before he even got to the presentation on the CF drug VX-770, the chief scientific officer had begun to pack up his briefcase. Dr. Van Goor glanced down at his PowerPoint presentation on VX-770, saw thirty slides, and realized he needed to hurry. He skipped to the dramatic last slide in the deck, a video showing the real-time effect of VX-770 on chloride clearance in the cell. First chloride is seen stuck inside the cell, with dry and hardened mucus on top of the cell; then VX-770 goes in, chloride goes out, and the mucus thins out visibly, bringing the cell back to life. The chief scientific officer put down his briefcase and started to ask questions about this new drug.

Vertex was still far from sold. Bob Beall, in a meeting with Vertex president Josh Borger, highlighted the intangibles—how both the scientists who had come over from Aurora and those at the CF Foundation were passionate about this project. He also emphasized that considerable sums of money from the CF Foundation had already been pledged, that there would be tremendous support for clinical trials in the form of a robust patient registry, as well as intellectual support from scientists throughout the country.

Fortunately for all, Vertex made the decision to stay in the CF game. There was potential in this new class of drugs, and in some ways, working with a foundation was easier than working with typical money-lenders, who applied constant pressure to produce a profit. The foundation was focused on one thing—finding drugs to bring to market to improve lives. Profit was secondary. Vertex decided to give this new model of venture philanthropy a chance.

Paul Negulescu and his team continued working on VX-770, and after some promising outcomes in actual human lung cells, they brought the drug to patients. Their hope was that VX-770 would improve some symptoms, slow the progression of disease, and be generally well tolerated. Instead, in a phase II trial published in 2010, involving twenty patients with the appropriate genetic defect, there was an 8.7 percent increase in lung function, a completely unexpected improvement. A larger follow-up phase III trial was published in the September 2011 issue of the *New England Journal of Medicine*. The results were again spectacular. Patients saw a 10 percent increase in lung function, an increase in weight, and symptomatic improvement.[13] A 10 percent increase in lung function, when CF patients were used to losing function every year, could make an immense difference in exercise capacity and overall well-being. These were unheard-of results in such a devastating and relentless disease.

Barely fourteen years had passed from concept to small-molecular drug discovery to successful clinical trial. No other disease this complex, and this deadly, had seen so much progress so quickly. Almost as importantly, the "proof of concept" moment had arrived for small-molecule medicine in CF. There was elation at the CF Foundation, but also cause for concern. The hike up Mount Everest had progressed significantly, but only for a small percentage of the patients. Ninety-six percent of them, those without the specific mutation helped by VX-770 (now known as ivacaftor), were still stuck closer to base camp than to the top.

The next small-molecule pill Vertex tested was Orkambi, a combination pill made up of the previously approved ivacaftor and the new drug lumacaftor. Targeted against the mutation delF508—the one Tsui and Collins had found—which is the most common genetic mutation in CF, it had the potential to cover 50 percent of patients. Its success was certainly not predicted from the odds, since some 90 percent of drugs in late-stage clinical trials never make it through to patients, either because they are ineffective, are not well tolerated, or have side effects that go unseen prior to human trials. Remarkably, the trial for Orkambi was a success, and the FDA granted approval on July 2, 2015.

Because we have about fifty patients with this mutation at our CF center in Philadelphia, prescriptions flew out of our office to the pharmacies. Thousands of prescriptions went out from hundreds of CF centers around the country. It was truly a historic day for about fifteen thousand patients: for the first time they had a pill in their hand that was actually modifying the disease and not just clearing and suppressing mucus.

Enthusiasm was tempered slightly, since the data was not as robust as it had been for ivacaftor and the G551D mutation; the study, published in 2015 in the *New England Journal of Medicine*, showed an average increase of 2.8 percent in lung function.[14] Fortunately, on the heels of the ivacaftor and Orkambi studies, new phase III studies started on another Vertex medicine, Symdeko. This drug also targeted the 50 percent of patients with the delF508 mutation, and results published in the *New England Journal of Medicine* in November 2017 demonstrated a slightly more robust response than Orkambi had had, with a 4 percent increase in lung function.[15] Despite the success and subsequent FDA approval of Symdeko in February 2018, some frustration remained, because the drugs Orkambi and Symdeko were only modestly effective, and still only about half of CF patients were eligible for a protein modulator.

The next big leap forward with protein modulator therapy came in 2019, with the published results of Trikafta, a pill that contains three protein modulators, all working in synergy. The study, published in the *New England Journal of Medicine*, showed an average increase of 14 percent in lung function, over three times the result seen with Symdeko.[16] Just as important, with its superior efficacy, the drug is powerful enough to work on many of those CF patients with mutations that caused more severe defects in the protein. With FDA approval of Trikafta in October 2019, the percentage of CF patients eligible jumped from 55 percent to almost 95 percent.

The stories coming in of patients on Trikafta are both disease and life altering. Patients who were watching their names travel to the top of transplant lists are coming off those lists. Patients who normally booked three or four trips to the hospital each year to manage their exacerbations are instead booking reservations for vacations. Blood coughed up on a daily basis has evaporated. Instead of waking up three or four times a night to cough out mucus, patients are having the strange experience of sleeping through the night and waking up refreshed. One patient commented, "It's been transformational. My cough is basically gone. I am able to walk without hacking. My lungs feel clear." Another, a middle-aged woman, gained fifteen pounds and remarked, "I finally have a butt! For the first time in my life! I am so happy!"[17]

Now that some 95 percent of CF patients are being treated with a protein modulator, their life expectancy is increasing. At the 2017 North American CF conference, Dr. Preston Campbell, head of the CF Foundation after Bob Beall's retirement in 2015, amazed everybody in attendance by announcing at the invocation that the estimated average lifespan for a patient born with CF had jumped from forty-one years to forty-seven years in the previous year, the single biggest increase in decades.

Even with the success of the FDA-approved protein modulators, the CF Foundation is not content. Gene therapy, once left for dead, is being revived with a different system. Just as viruses have been battling with humans for millions of years, they have been fighting with bacteria for millions of years as well. As part of this combat, bacteria developed an enzyme, called CRISPR, that can both cut up invading viral DNA and insert normal DNA back in. The CF community envisions using the CRISPR enzyme to cut out the defective portion of the genetic code in CF patients, while at the same time giving the CRISPR system a template for a normal gene to insert once the defective gene has been eliminated. If this could occur at the stem-cell level, all subsequently produced cells of the lung would have a normal CF gene.

Beyond protein modulators and gene therapy, there is an effort in CF research to improve affected cells' RNA, the messenger material that is made from the template DNA and directly results in protein formation. Some of this work is being done in Lexington, Massachusetts, where, after selling their portion of the rights to ivacaftor to Vertex, the CF Foundation was able to open its own state-of-the-art laboratory in 2016, with twenty-five full-time employees. Their focus is on treatments for the 5 percent who will not benefit from protein modulators because of the severity of their defective protein. Therapies involving airway clearance, antibiotics, nutrition, and the pancreas are all going forward. Michael Boyle, the current head of the CF Foundation, is adamant that one day, in the not-too-distant future CF will stand for "Cure Found," and patients will be saying to their friends, "I used to have CF."

Chapter 15

Cystic Fibrosis, the Most Heartbreaking Lung Disease

No one understands the importance of the breath better than Sarah Murnaghan and her family. Born in 2003, Sarah appeared to be healthy at birth, and even though she was born premature, at thirty-six weeks, Sarah was an average size and weight. At home though, her mother, Janet, soon realized there was something very wrong, as Sarah wouldn't eat, wouldn't stop crying, and was unable to gain weight normally. Even after she drank milk, her diaper was almost never wet. The pediatrician told Janet that everything was okay, but even as a new mom, Janet knew something wasn't right. And she was correct. After eighteen agonizing months of prodding physicians, Sarah proved to have cystic fibrosis.

Janet and her husband, Fran, were in shock. Things that happened to other people had happened to them. The survival statistics for CF weren't encouraging, but neither were they completely devastating. The average lifespan for children with the disease at that time was twenty-eight years, and it had gone up ten years in a generation.

Even with effective medicines, CF requires a tremendous amount of work, most of it directed at the lungs, the organ responsible for 90 percent of the mortality associated with the disease. Cystic fibrosis patients make too much mucus, which gets stuck deep in the lungs, providing a fertile breeding ground for bacteria. Infections and lung destruction follow in a mutually reinforcing cycle of dysfunction.

To help prevent or, more realistically, delay damage to the lung tissue, patients inhale multiple medications every day to break up the mucus. After taking the medicines, they must expectorate the secretions, either with vigorous coughing or with chest physical therapy, in which a parent pounds on the child's back to agitate the mucus and move it out of the airways. The work is intense and usually requires an hour in the morning and another hour at night, even with the advent of protein modulators. And like Sisyphus's task of rolling the stone up the hill, it must be done every day.

Janet and Fran did Sarah's airway clearance religiously, gave her the inhaled antibiotics, and brought her to her doctor at the Children's Hospital of Philadelphia every three months. Most CF patients don't manifest debilitating symptoms until they are in their twenties or, if they're lucky, their thirties. But the late diagnosis at eighteen months had cost Sarah, and one of her lungs was already permanently damaged. By the time Sarah was seven, she was going into the hospital every few months for intravenous antibiotics. Her lung function would stabilize, and she and her parents would breathe a sigh of relief. But then another infection would develop, along with diabetes, osteoporosis, asthma, and malnutrition. Sarah missed a lot of the second grade, and with the threat of viruses and bacteria at school, she began homeschooling. She never complained and never felt sorry for herself, but it broke her parents' hearts that she couldn't lead a normal life.

Then, when she was nine, Sarah ended up in the hospital and could not return home. Her lung function was at 30 percent, and the bacteria that colonize every CF patient had become completely resistant to antibiotics. She needed to be put on oxygen, and she struggled to walk. Weight loss became an issue, and she dwindled to fifty pounds. Her parents saw the happy daughter they once knew start to slip away, replaced by a girl robbed of her childhood.

With this level of illness, Sarah couldn't go on much longer. The time to change course in the management of her disease had come, and Dr. Howard Panitch had to break the news to the family. The Murnaghans trusted Dr. Panitch unequivocally. He had been a cystic fibrosis doctor for thirty years, and Sarah's physician since her diagnosis. His refusal to accept average outcomes matched Janet's approach perfectly from day one. Even so, when he told Janet and Fran that Sarah would need a lung transplant to survive, they were stunned. Nonetheless, they both accepted his assessment and were soon introduced to Dr. Samuel Goldfarb, the medical lung transplant doctor.[1]

What Dr. Goldfarb told the Murnaghans next stunned them even more: while Sarah needed a lung transplant, it was doubtful she would live long enough to get one because she was subject to what was known in the lung transplant world as the "under-twelve rule." Adults are listed for lung transplants in an order based on their level of illness. But because of a lack of data on how pediatric patients should be prioritized, children under the age of twelve were, in 2012, left in an antiquated system in which patients simply got in line. Those who were able to stay alive long enough would get their new lungs; however, because of the paucity of pediatric lung donors, many children died while on the waiting list. A little girl as sick as Sarah would almost certainly be one of those to die. There was almost no chance she would last long enough to get a new set of lungs.

Janet Murnaghan was incredulous when she learned how lung allocation was organized. To her, patients at the greatest risk of dying should get organs first. The system for adults had shifted away from the first-come-first-served method in 2005. Assigning organs based on need was an ethical, equitable principle that nobody would contest. But the under-twelve rule was still in place, and it would not be easily changed.

With Sarah stuck on a long waiting list serviced by a scarcity of donors and unable to get into the more sensible adult system, Janet made a simple but powerful statement: "The only thing standing between my daughter living and my daughter dying is the fact that she's ten and not twelve. That's unreal to me."[2]

Janet's statement, and her subsequent fight on social media, in the press, and ultimately in the courtroom, brought to public attention the rationing of organs, patients dying on waiting lists, and, most important, the fate of a little girl with a lethal genetic illness and scarred lungs. Many people would weigh in, including the head of the US Department of Health and Human Services, federal judges, doctors, ethicists, the entire lung transplant community, and people throughout America and the world. In the end, the outcome left some people satisfied, and others completely dissatisfied. One thing did remain steadily clear throughout the whole ordeal: a little girl, her mother, and her family, refused to give up in their quest to restore the breath of life to one from whom it had so unfairly been taken away.

I attended my first transplant lecture during year one of my lung medicine fellowship. The speaker's first slide was simple: a glass of water exactly half full. Over the course of the next hour, he explained the meaning of the image as it related to lung transplants. The glass half full gave patients a new lease on life—a new set of lungs and a vastly improved breathing experience. The glass half empty was the stark reality that lung transplants are fraught with potential life-threatening problems, such as infection and rejection.

Each lung transplant candidate must weigh the benefits and drawbacks of a transplant. For Dr. Panitch, Sarah, and her parents, the conversation was straightforward. Sarah was stuck in the hospital on potent antibiotics with an oxygen mask strapped to her face. Pain was a big issue from compression fractures in her back due to long-term steroid use. At night she played monopoly with her mother, and before bed they would turn up the air conditioning to simulate the breeze of a beach, the air rocking the paper lanterns on the ceiling. Her friends went to school every day, while she went about the business of trying to keep herself alive. Sarah and her family were forced to see lung transplant as a glass half full, and to take their chances with infection and rejection.

Once the transplant decision had been made, Janet's crusade to change the rules began. The issue with lung transplants—and with all transplanted organs, she discovered—is the simple fact that there are more people in need of an organ than there are healthy organs available. This creates a system in which patients die on waiting lists. For the years between 2015 and 2019, 935 patients died on the lung waiting list, an average of 187 patients per year, or one every 47 hours. The statistics for other organs are similar. In 2019 alone, 5,445 people died on waiting lists for all types of organ transplants. From 1995 through 2019, 166,223 patients died awaiting a transplant.[3]

When resources are limited, a basic ethical tenet is that those who need the resource the most, namely those who are the sickest, should be prioritized. The system for prioritizing adult lung transplants changed in 2005 because of an order from the Department of Health and Human Services, known as the "final rule," which states that lung transplant allocation must take a patient's level of illness into account when prioritizing their position on the waiting list. But level of illness was not the only criterion. If only the sickest patients received transplants, many of them would never recover, and the organs would essentially go to waste. A measure of a patient's ability to survive after a transplant had to be added into the calculation.

Balancing level of illness against the ability to bounce back from a big surgery is difficult in lung transplants. Given the overall poor outcomes and ubiquity of complications, transplanting lungs is a high-wire act compared to transplanting other organs. Assessing a patient's likelihood of dying soon from their lung disease versus their ability to live after a transplant was considered impossible. Thus, the original system of getting in line and waiting for new lungs remained until 2005, long after it had been abandoned in liver and kidney transplants.

Using data from previous lung transplants, physicians and statisticians developed a new system called the Lung Allocation Score (LAS). Each patient awaiting lungs received a score between 0 and 100, with those scoring closer to 100 being highest priority on the waiting list. The equation to calculate the score was based on how likely the patient was to die while on the waiting list matched with how likely the patient was to be alive one year after the transplant. These two calculations helped balance justice—the principle that everybody should have equal access to resources—with utility—the reality that a limited resource needs to be allocated to those who will benefit the most from it.

The system still wasn't perfect, as the calculations were little more than educated guesses, and unlike the system used to allocate liver transplants, the equations used for lung transplants were not prospectively validated. Prospective validation is a big deal in medicine. The best way to derive the calculations for balancing need of a transplant with likelihood to survive a transplant would be to create two groups of patients, one that received transplants and one that did not, and then compare how long they lived. This obviously could not be undertaken ethically, since patients waiting for a lung transplant could not be denied an organ just so formulas could be derived. In contrast, liver and kidney formulas can be more accurate because their absolute numbers are so much larger that statisticians have more data to work with.

Fortunately, the Lung Allocation Score system worked, despite its potential flaws. The number of deaths on the waiting list plummeted in the years after the new system was implemented, from an average of more than four hundred to two hundred per year. Almost as important, there was no big change in one-year mortality, or even five-year mortality. The doctors were performing transplants on the sicker patients, saving lives on the waiting list, and not putting organs to waste.[4]

However, the issue of children under twelve remained a problem well beyond 2005. Originally, doctors believed that size disparity would prevent an adult set of lungs being successfully transplanted into a child. Also, an LAS number for children under twelve could not be easily formulated since there was so little data for statisticians to draw on: between 1990 and 2001, seven thousand adult lung transplants had taken place in the United States, yet only four hundred transplants had been performed on pediatric patients in the same time period. Pediatric lung transplants were also done to cure different diseases than adult lung transplants, conditions like congenital protein deficiencies and heart diseases that also affect the lungs. Different diseases do not behave the same way under transplant conditions. The LAS formulas could not be easily tweaked to fit the younger population because of this disparity.

Given the small number of under-twelve transplants, statisticians felt they did not have enough data to devise equations to answer the two dispositive questions: Who is going to die awaiting a transplant, and who is going to benefit the most postoperatively? So for eight years after 2005, the dual allocation system, however flawed, remained in place. Those twelve and over could be eligible for adult lungs, while those under 12 remained in their own donor pool, ranked by level of respiratory failure, blood type, and time spent on the waiting list—until Janet Murnaghan started questioning the system's ethics.

The first thing Janet recognized was that the under-twelve rule was a completely arbitrary designation. If organ or body size was an issue, then age was not really important: some ten-year-olds were as big as thirteen-year-olds, or even eighteen-year-olds. The reverse was also common. Janet also found out that size was not necessarily an issue at all. Each human lung has three lobes on the right and two on the left. These lobes can be separated and trimmed out as needed. Several published reports concluded that an adult set of lungs could be cut to fit a child, with good outcomes.

A final source of frustration for Janet was that Sarah had a disease for which an LAS predictive model existed. A claim could not be made that her disease was unique to pediatrics. Cystic fibrosis was a common reason for adult transplant, so basic expectations existed of what to anticipate at certain levels of illness. There was no evidence that the predictive models for dying on the waiting list or being alive one year after transplant would be any different for kids with CF than for adults.

Janet and her family felt they could deliver more than enough proof that the under-twelve rule did not fulfill the transplant principle of justice. Janet also felt that if doubt remained as to what the policy should be, it should be rewritten to err on the side of being inclusive rather than exclusive. The rule change would affect few patients, but at the same time benefit very sick children. But not everybody agreed. Some people, in fact, vehemently disagreed with Janet, including some of the leading transplant experts in the world. They argued that the systems had been designed with the best available evidence at the time, and rules shouldn't be changed based on single cases for those with the loudest voice or deepest pockets.[5]

With their daughter slipping away, Janet and Fran saw two options. Sarah had been listed for a transplant for over a year, and none had materialized. Now was the time to either let nature take its course or investigate the possibility of trying to get on the adult list for lungs. They asked Sarah if she wanted it all to stop. Fully aware of her situation, Sarah was unequivocal: "I will never give up, ever, so don't you give up on me!"[6]

With a clear answer, Janet asked the transplant surgeon and medical doctor whether they would be willing to do a lobar transplant on Sarah, that is, to cut the lungs from an adult donor into lobes, extract the two best, and make them fit into Sarah. Their answer was unambiguous: not only would they be willing to do it, but based on published reports, they thought the transplant would be successful. With newer immunosuppressant drugs and care, and increased understanding of rejection issues, lung transplant patients were living longer than ever before. Even more important for Sarah, within the group of transplant patients, CF patients excelled. The fifty percent survival rate for all patients had risen to somewhere between six and seven years. For CF patients as a subgroup, it was about seven and a half years in 2013.

Now Sarah and her team just needed to get that blessed organ, and the medical community needed to figure out if it was ethical to give Sarah one.

In 2013, Sarah Murnaghan hoped to receive new lungs, exactly thirty years after Joel Cooper had accomplished the first successful transplant. A former public relations executive, her mother determined to put pressure on the United Network for Organ Sharing (UNOS), the nonprofit group that manages the nuts and bolts of every organ transplant in the country through the Organ Procurement Transplantation Network (OPTN). With advice from the transplant doctors, surgeons, ethicists, and epidemiologists, UNOS and the OPTN make all the rules for how organs are distributed in the United States.

Janet crafted a letter to her friends, telling them of Sarah's dire condition, and how the only way to save her was for her to get on the adult list in the appropriate place as dictated by her level of illness. She hit "send" on the e-mail at 10:00 p.m. on Friday, May 24, 2013, then closed her computer and got into bed with Sarah, among the tubes and wires and beeping monitors, Fran asleep in the cot next to the bed.

The next morning Janet logged onto her computer, hopeful for a few replies—instead she had hundreds of messages, and a behind-the-scenes plan for a public relations blitz had developed overnight. The *Philadelphia Inquirer* ran the first story, and then the family was contacted by CNN. A crew happened to be in town over the Memorial Day weekend, and was sent to Sarah's hospital. The piece aired on CNN on May 27, starting with a video of Sarah, sitting on the edge of her hospital bed with an oxygen mask strapped to her face, singing "Twinkle, Twinkle, Little Star" while striking the notes on a xylophone.[7] Entire elementary schools and nursing homes around the country wrote letters of support.

The OPTN, however, did not budge, and their physician representative replied: "It tugs at my heart. It's not a perfect system. There is no perfect system. It's the best we can do right now. If I change the system to give Sarah an advantage, there's another patient, very likely an adolescent, who then gets a disadvantage. We've built a system that tries to be as fair to everyone as possible."[8]

In response, Janet and Team Sarah started a petition at change. org, firmly believing there was no medical reason why Sarah or any other youth under twelve and in the same circumstance shouldn't be allowed to get in line for an adult organ. Five hundred people signed it. Then one thousand. Then one hundred thousand. Then four hundred thousand. Each signature generated an e-mail to the president of the OPTN. His inbox crashed after forty-eight thousand.

Another turning point was reached when senator Pat Toomey and then-congressman Pat Meehan, of Pennsylvania, joined the cause and reached out to Kathleen Sebelius, secretary of Health and Human Services. Janet and Sebelius talked on the phone, but Sebelius would agree only to review the policy, which would take months and be too late to help Sarah.

The arguments that Sebelius used to come to her decision had obvious merit. The lung transplant rules were set up in a transparent manner with the best evidence available at the time. If the injustices to the pediatric population were rectified, then injustices would potentially be done to another group. With a scarcity of organs, an adult set of lungs given to a pediatric patient could mean one more death on the adult waiting list. And even though ten case reports existed of pediatric patients doing well with adult lungs, thousands of cases supported the knowledge that adults do well with adult lungs.

Heartbroken over Sebelius's stance, the family as a last resort took their case to the courts. The hearing was held on June 5, 2013, and was overseen by the federal judge for the Eastern District of Pennsylvania, Michael Baylson. The star witness was Dr. Samuel Goldfarb from the Children's Hospital of Philadelphia, their lung transplant doctor. Judge Baylson got right to the point in his questioning: Would Sarah live as long as an adult after the transplant? Would her quality of life be good? Would adult lungs work on Sarah? Dr. Goldfarb answered each question in the affirmative. He explained to the judge that the under-twelve cutoff was arbitrary and not rooted in any science; it could easily have been a different number.

In a stunning rebuke to the OPTN and Kathleen Sebelius, Judge Baylson ordered a temporary suspension of the under-twelve rule, stating that it "discriminates against children and serves no purpose, is arbitrary, capricious and an abuse of discretion."[9] The Murnaghans were euphoric. Sarah's transplant doctors could now calculate an LAS score that actually meant something. Given her level of illness, Sarah's score put her at the top of the list for adult lungs.

An adult donor became available a week later. It was just in time, as Sarah had been sedated and placed on a mechanical ventilator a few days earlier, her lungs finally giving out under the weight of relentless infections, mucus, and fevers. The surgeons at Children's Hospital of Philadelphia trimmed down the adult lungs, took Sarah's old lungs out, put the two halves into her chest, sewed up her vessels, and released the blood and air into the donor lungs. The surgery had started at noon and had extended into the early evening. Everything went according to plan.

But when Sarah was wheeled out of the operating room, things started to go wrong almost immediately. The doctors couldn't remove her breathing tube because the level of oxygen in her blood was much lower than expected. A chest X-ray was done, and it confirmed what everybody had dreaded: the lungs were failing, in spectacular fashion, and given the amount of inflammation seen on the chest X-ray, and the level of oxygen in her blood, it was clear that nothing could turn them around. The only option was to try for another transplant. In the meantime, the ventilator wouldn't be enough to keep Sarah alive. The doctors were forced to put Sarah's heart and lungs on a bypass machine, and keep her completely paralyzed, a very short-term stopgap prior to a second transplant. This was also the last try; precedent existed for one retransplantation, but not for a second.

Dr. Goldfarb and the transplant physicians calculated another LAS for Sarah to get her back on the list for a retransplant, the number being extremely high again because of her catastrophic level of illness. Three days later, the doctors got a call from the OPTN: a set of lungs had become available, but they contained a clear area of pneumonia. They could try to wait for better lungs, or they could cut away the pneumonia and treat the remainder of the lungs with high-powered antibiotics. In consultation with the Murnaghans, they choose to take the lungs with pneumonia.

For the second time in a week, a greatly diminished Sarah went back into surgery. The surgeons cut out the infected portion of the donor lungs and began the surgery again. Afterward, they didn't try to remove the oxygen tube right away, or even close up Sarah's chest; her lungs expanded and contracted in full view underneath a thin, transparent protective film, her heart beating in the middle.

Sarah stayed on the ventilator for the next week while the team balanced the inherent opposites of fighting an infection with antibiotics and warding off rejection with immunosuppression. The swelling in her body slowly improved, and they began weaning her off of the ventilator. A week later, they took Sarah back to the OR and closed her chest. A day after that, they woke her up for the first time, and she somehow summoned the energy to sit on the edge of the bed. The next day, she sat in a chair, and the day after that she painted a picture while sitting in that chair. She was home at the end of August—on a ventilator, but home nonetheless.

* * *

Sarah stayed in the media for a while after her second transplant, but like many stories that have captured the front page, hers faded out, and she was left alone to do the work of rehabilitation. I followed her story for a while but then, like most, lost track of it. Sarah left the headlines, and discussions before conferences and after rounds in the hospital strayed to other topics.

In February 2014, about nine months after Sarah's transplants, I attended a dinner to honor unsung heroes in cystic fibrosis. Every year, each cystic fibrosis center in the Philadelphia area nominates a member of a care team as well as a patient. I was the physician representative from my center. The dinner landed on Valentine's Day that year, and more than five hundred people were in attendance. During the dinner, Dr. Howard Panitch stepped to the microphone to give out the Shining Star Award to the person with cystic fibrosis who had displayed the most courage and overcome the most adversity in the previous year. That year the award was going to Sarah Murnaghan.

"As a physician in an academic medical center, I am used to my role as a teacher, both to other medical professionals and to families," Dr. Panitch said. "Occasionally, though, we are taught special lessons from our patients that profoundly influence how we care for others, or how we aspire to conduct ourselves. I am humbled to be one of Sarah's students, and hope that I can meet my own challenges with the same grace and determination that she demonstrates every day. The Shining Star Award recognizes a person with CF who strives to live life to the fullest and who overcomes the many obstacles that CF presents. I can think of no one more deserving of this award than Sarah Murnaghan. Please join me in congratulating her on receiving this year's Shining Star Award."[10]

In a purple sparkle dress and black shoes, Sarah stood up out of her wheelchair and approached the microphone. Her voice was a whisper, but her quiet, simple words resonated. We all stood transfixed by her.

"Special thanks to my doctor, and there are a few other things I'd like to share. I have a really strong family, the best in the world, my mom and dad never left my side, my brother and sister and cousins cheered me on. Aunts, uncles, and grandparents, you name it, there is no give-up with this bunch. It is because of them that I knew that I wanted to live. And the one thing that I needed to fight is the most powerful weapon: God gave me bravery. The reason I survived two transplants, wasn't just me, it was the bravery in me. And knowing that my family would never give up on me. And I know all CF kids and kids with other diseases, too, have that bravery in them. So my advice, whatever you are dealing with, young or old, is to be yourself. Look deep inside yourself for that bravery, and the reason to fight. For me it was my family. It may not seem to be there at first, but I promise you it's there, and it will take you to the impossible if you just believe in yourself. Thank you."[11]

The audience was quiet for a moment, and then it burst into applause. Some of Sarah's friends from school had come with her, and they now swarmed her. A feeling of electricity was in the air. The politics, the ethics, all melted away for this moment as we all pondered the lessons Sarah had just imparted to us, and the fact that this girl, who had been through so much, was, at the age of eleven, teaching us about the meaning of life.

Joel Cooper once heard in synagogue, in reference to the parting of the Red Sea by Moses, that a miracle is an event that leaves you with an abiding sense of astonishment. That is the exact feeling I had that night after witnessing Sarah's speech.

In June 2014, on the one-year anniversary of Sarah's transplant, the lung transplant committee finally made official the rule that allowed Sarah to obtain an adult set of lungs. Other children are now free to apply for a set of adult lungs, and over twenty have, bearing out Janet's argument—that the policy change was not just about Sarah.

Seven years after her transplants, Sarah continues to make impressive strides. She went back to school, jumped into swimming lessons, and even joined a competitive team. Every lung transplant patient takes things day by day, but by any standard Sarah is making the most out of her days. Her life is still not easy or simple—she takes multiple medications, goes to see her doctors regularly, and is at extremely high risk for organ rejection, as well as for strange and nasty infections. The good thing is that she no longer needs to do airway clearance, since her new lungs have normal amounts of CF protein (which is also why protein modulators won't do her lungs any good). All things considered, a better outcome for a lung transplant patient is hard to imagine. When Dr. Goldfarb got in front of Judge Baylson and told him he believed Sarah could do very well, he was right. Her quality of life has improved immensely, one of the most important outcomes that can be hoped for. Despite all the controversy, lung transplant remains an example of the very best that we are capable of in medicine.

While the ethical issues surrounding the process have not all been resolved—fierce debate about changing the under-twelve rule continues in the lung transplant literature—one positive outcome has been the widespread press coverage for the organ, so frequently ignored, that provides the breath of life.[12] As Sarah's case makes clear, life without healthy lungs is cruel, but with them it is glorious.

In 2011, Shelley Dobson was thirty years old and had lived with cystic fibrosis every day of her life. When she came into the world, CF was with her, and when she leaves the world, its scars on her organs will be obvious. To keep her lungs healthy, Shelley has to take several inhaled medicines, usually twice a day. They go in through a nebulizer, and each dose takes ten to twenty minutes to inhale. Some of them are irritating, and result in a violent cough that can turn Shelley purple.

After inhaling the medicines, Shelley does intensive airway clearance, the most important thing. Airway clearance, as described in the previous chapter, involves some type of physical agitation, followed by deep inspiration and coughing to blast the mucus out of the chest. When Shelley was a child, her mother would pound on her back to dislodge the mucus, turning Shelley over into different positions to focus on different parts of the lung. Later, Shelley used the vibrating vest, strapping it on tightly for thirty minutes at a time.

Many CF patients are diagnosed right at birth, but Shelley wasn't. She is African American, and CF is almost exclusively a Caucasian disease; she also had two perfectly healthy older siblings. At three months, though, her mother came home from work, and the babysitter reported that Shelley hadn't had a wet diaper all day. Concerned about how Shelley looked, her mother brought her to the emergency room, where she was simply told Shelley's levels of potassium and sodium were completely off. Her mother left it at that, but when this happened a second time a few months later, and then a third time, she refused to leave the ER. Shelley ended up staying in the hospital for thirty days, and was ultimately diagnosed by Dr. Howard Panitch, who would be her doctor for the next two decades.

Although her mother was told Shelley's life expectancy was twelve years, Shelley reached the age of thirty through tenacious hard work, and around this time I became her physician. Thirty years is a long time to live with a demanding chronic illness. The burden of her therapies, of the constant struggle to keep her lungs healthy, to keep her breathing going, was starting to weigh on her.

Shelley had also complicated her life. Five years earlier, against the advice of her doctors, she had had a son, now a beautiful boy with a million-dollar smile. He was a kid who would run up to you and hug you when you walked into the room, even before you had a chance to say hello. And like all mothers, Shelley was troubled from time to time by thoughts of what her child would do without her, and also what she would do without her child. Those thoughts were becoming louder as her lungs began to show the burden of three decades of cystic fibrosis.

The biggest sign I got that something was wrong appeared on a Friday afternoon in early February. It was cold outside, and the snow and ice had built up during the Philadelphia winter. Shelley sat in the chaotic emergency room in the downtown city hospital where I worked, with machines dinging, nurses scurrying around drawing blood, and overhead pages blaring all around us.

Shelley had come in because she had been coughing up a lot of blood, not just the usual streaks in a napkin. I walked into her room and squeezed her hand. She was too tall for the hospital bed and looked awkward lying there. Her long, thin arms dangled over the railings, and an IV pole hung ominously above her, dripping in an antibiotic. She always smiled an affectionate, inviting smile, just like her son. There was no sourness in Shelley, only warmth. But today the smiles were hard to come by. She didn't look up at me, which was unusual for her. Her husband held her hand tightly through the railing on the other side of the hospital bed as I asked the usual doctor questions: When it had started? What else, if anything, was going on—any fevers or chills, chest pain or tightness? When she responded, I could detect a hint of resignation in her voice for the first time. "Dr. Stephen," she finally said to me, in her quiet, dignified way, the pinging of the monitors fading into the background as she spoke. "I'm really scared for one of the first times in my life. I know what I need to do, but I'm scared."

After her diagnosis at six months of age, Shelley's parents settled into the work of taking care of a child with a very serious illness. With the help of their doctors and team at St. Christopher's Hospital for Children in Philadelphia, Shelley's parents watched with wonder as their tiny baby grew slowly but surely into a typical little girl, who loved to play and get dressed up and dance and sing. Her mother helped her with her therapies, and overall she did very well. They were regimented about all the inhalers and the airway clearance, and Shelley proved wrong the doctor who predicted she wouldn't live past twelve.

About once a year, however, Shelley found herself gripped by an acute exacerbation, something common in CF patients, which occurs when the bacteria that have chronically colonized the damaged airways multiply. The symptoms are not friendly, with fever, shortness of breath, fatigue, and weight loss knocking patients down in an instant. These bacteria must be quickly controlled, and this usually requires intravenous antibiotics for two weeks. Many CF patients suffer two or three of these episodes a year, learning to be in tune to symptoms so they can get themselves on IV antibiotics before falling off a cliff.

Starting when Shelley was ten, every March an exacerbation would send her to the hospital. Back then, CF patients could visit each other freely, and Shelley got to know the other kids with CF. They would trade music on their Walkmans, and at night put on plays for their parents, turning off the lights in the hospital room and using flashlights to illuminate their "stage." Shelley's mother stayed with her every night, leaving only in the morning to go to work.

In the early 1990s, two new medicines were approved for use in CF. The first of these was the inhaled antibiotic tobramycin. Doctors believed that CF patients could achieve much higher levels of effectiveness and avoid toxicity to the other organs if the medication was delivered straight to the lungs. When given too long intravenously, it created potential toxicity to the kidneys and the ear. However, this risk was obviated with the inhaled therapy, and patients could take it every day for months at a time.

The reason tobramycin was such an important medicine was that most of the patients with CF had lungs chronically colonized with a nasty bacterium, *Pseudomonas aeruginosa*. And like an unwelcome squatter, once *Pseudomonas* hit the lungs of a CF patient, it was almost impossible to eradicate. Tobramycin made no promises of eradication, but it could keep the bacterium under control. It worked as promised, and a 1997 trial showed it led to a 10 percent improvement in lung function for the average CF patient.[13] Ten percent for a CF patient could mean the margin between one hospitalization a year and four, between being able to go running for half an hour or being stuck at home coughing.

The other medicine was Pulmozyme, also delivered to the lungs. Pulmozyme is not an antibiotic, but a dicer of DNA. *Pseudomonas* causes a massive inflammatory response, and huge numbers of white blood cells influx into the lungs to combat it. Pulmozyme works by cutting up the DNA of the dead white blood cells, effectively sweeping dead autumn leaves off the ground. This wasn't anything fancy like gene therapy, but in a large trial it was also shown to give CF patients a modest increase in lung function.[14] For Shelley, these two medicines helped stabilize her lungs. They didn't stop the exacerbations, but they generally increased her exercise tolerance and reduced daily mucus production.

By the time Shelley entered high school, she had grown tall and beautiful, one of the tallest students in her school. The new inhaled therapies did help, but exacerbations kept coming on a yearly basis. An ominous moment occurred early in high school, when she coughed up blood for the first time. It scared her, as it would anybody, but even more because she remembered watching the movie *Alex: The Life of a Child*, a documentary about the life of Alex DeFord, daughter of sportswriter Frank DeFord who was also afflicted with CF. Alex coughing up blood into the white basin of a sink was an image Shelley never forgot. Alex would die at the end of the movie, all of eight years old. It was only natural Shelley believed she was on the same path, albeit a few years later in her life.

Shelley endured, finished high school, and focused on getting an advanced degree in nursing. Having cystic fibrosis and having had exacerbations, she knew the importance of nursing staff. Doctors are realistically in the room for about five minutes every day. Nurses are in and out of the room multiple times over a twelve-hour shift.

In college, Shelley met a man, Franklin, and they fell in love. He understood her, and also understood her disease. After dating for a while, they got married, and with him joining the Coast Guard, they moved from Philadelphia to Virginia. This was a big change for Shelley at age twenty-four, as her friends and family, especially her mom and dad, had been her support.

Shelley was equally nervous about being away from her doctors and transitioning into the military health system. Used to major academic medical centers, and with her persistent intermittent exacerbations, she had misgivings about her new doctors' ability to effectively treat her. Fortunately, her misgivings were misplaced. The military doctors ended up being very good and attentive, and Shelley actually improved with her disease and was able to do something many CF physicians counseled against, believing it would be too much stress, enough to kill a woman with CF. She got pregnant.

Shelley, though, was ecstatic about her pregnancy. Always refusing to be defined by her disease, she felt having a baby was a natural progression in her life. Many old-school CF physicians would have scolded her for it, but Shelley knew her body better than anybody. With her optimism, work ethic, and support, she was quite convinced she was going to have a happy, healthy child, and that her own health would not suffer for it.

Shelley gave birth to a beautiful, happy baby boy. Her husband, Franklin, hadn't wanted to get tested for CF, and it seemed like an okay decision. For their son, Jason, to have CF, he would have to get a CF gene from both his mother and his father. Like Shelley, Franklin was African American, and the chance that he could be a carrier of a cystic fibrosis gene was low. The carrier rate for somebody of European descent is about 1 in 29. This drops to 1 in 65 for African Americans, or about 1.5 percent of an average population. But even if Franklin did have the mutation, there was only a 50 percent chance he would pass the gene along to his son.

In the end, it wouldn't have changed what they did. But Shelley was astounded when the doctors told her that Jason's newborn screening test was positive for CF (newborn screening for CF began in 1995 in some states, and not until 2010 in others). It was like lightning striking twice. She experienced a barrage of emotions as she realized he would likely have to go through the same things she did—the hospitalizations, the doctors' visits, the daily treatments. Ultimately though, the disease of CF paled in the face of what a human life meant, and she determined to keep him as healthy as she could, as her mother had done for her.

Jason did amazingly well from the start. The parallels between their lives were striking—both African Americans with CF, both doing treatments in unison, the two of them grabbing inhalers and strapping on their vibrating vests to get mucus out together.

But Shelley also noticed some very important differences. For as long as she could remember, and that included when she was young, she always had a little cough, and would bring up mucus intermittently, especially in the morning. Jason had no cough, even as he approached his sixth birthday. With the newer inhalers and vests, the disease had changed in a generation.

If things were going well for Jason, Shelley was having a rougher time. Things hadn't worked out in Virginia, and she had separated from her husband and moved back to Philadelphia. Franklin also moved back to Philadelphia, and he remained a huge supporter of both her and Jason, but they weren't meant to be together long term. Shelley, always driven, took up her studies to become a nurse again, but with work, cystic fibrosis, and a son—who also had CF—it was simply too much. She backed off on her studies, but still finished a degree as a medical assistant.

With so much going on, Shelley had some difficulty keeping her weight up. The exacerbations also came on worse than ever, often preceded by coughing up blood. By 2011, at the age of thirty, the antibiotics for her exacerbations were taking longer and longer to work, and instead of the usual two weeks of antibiotics, the treatment courses were stretching to three weeks.

Shelley had dealt with CF for thirty years, and despite all the treatments and doctors' visits, CF had remained separate from her life because it had never controlled her, had never stopped her from doing anything. Now, with the increased exacerbations, the blood coming up, the time in the hospital, her life was spiraling downward like never before. During this time, I would see Shelley in the emergency room at the hospital, each time a little sicker, coughing up a little more blood, and a little more frustrated than before.

In 2011, at our CF center in Philadelphia, we were hearing rumblings about new medicines on the horizon, protein modulators that could radically change how we treated cystic fibrosis. We also knew that only certain patients would qualify, and we would have to know all of our patient's genetic mutations to see if they were eligible.

Prior to the advent of protein modulators, an individual patient's genetic mutations had been an afterthought. CF was diagnosed with the sweat test, and knowing a patient's mutations was somewhat interesting but not essential. With the modulators on the horizon, a hunt went on in CF clinics throughout the world to determine which patients had which gene defects. Over one week in September 2011, I began each morning by pulling a few charts at a time from the bookcase housing the files of our some hundred patients, searching for elusive pieces of paper from five, ten, or even fifteen years before that told us which mutations each patient possessed.

The majority of the charts contained the information, usually on a faded yellow sheet of paper with an old type font. I put the patients' names and their mutations into a spreadsheet on my computer. When I saved the document, I saw that it took up twenty-five kilobytes of memory—a miniscule amount of space. But this spreadsheet was perhaps the most important document on my computer. It would soon dictate what medicines certain patients would be able to try.

Ivacaftor, the new drug from Vertex Pharmaceuticals, was approved by the FDA for use in patients with the G551D mutation on January 31, 2012. At my university, we did what hundreds of centers throughout the world were doing, checking our database to find which patients held this mutation. We had four, the precise number of patients predicted by statistics. And one of them was Shelley.

The medicine was so new, our computer system didn't have it in its database yet. So we found an old-fashioned paper prescription pad and I wrote out four prescriptions in long hand, one for each of our four patients with the matching mutation. When we saw Shelley in the clinic, I noted how she carefully folded it and put it in her purse.

We handed out the other three prescriptions, and we crossed our fingers that something good was going to happen. A month passed, and to our delight some extraordinary stories started filtering in. One CF patient told us about her dry mouth. For some reason CF had affected her salivary glands, and her mouth was chronically dry and cracked. She constantly popped hard candy to stimulate her glands. After her first dose of ivacaftor, she noticed a strange sensation on her tongue and in her cheeks. At first, she didn't recognize her own spit because it had been so long since she had produced any. She was so overtaken with joy, just to experience something normal like that again, that she started crying.

Other stories came in reporting that the medicine didn't just work on the CFTR protein in the lung, it worked everywhere in the body, including the pancreas and gastrointestinal tract. One of our patients had this good fortune. Not only did his lung function improve, but his scarred pancreas began working and he was able to stop taking his pancreatic enzymes before meals. For the first time, he was able to just sit down, eat, and digest as we all do. He was also able to get off his inhaled antibiotic, creating an extra half hour in his mornings and evenings.

For some reason, Shelley did not have any dramatic stories to tell, probably because her attitude was so positive to begin with. There is no negativity in her to combat, no suppressed demons. She always tells you how she feels and focuses on the positives in her life, her beautiful son, her work. She is so even-keeled that nothing could radically change her outlook. Not even ivacaftor could do that.

Eventually, though, as the months passed, the changes became noticeable. Shelley stopped coughing up blood, and her lung function crept up, along with her weight. The exacerbations ended, and when we saw her in clinic, the conversations were short, whereas before she had lingered to talk about this symptom or that issue.

Her son, Jason, was also able to go on ivacaftor. At conception, Shelley had given him only one of the two genes for CF, but we found out later she gave him the one that was now treatable. She told me they always took their pills together, every morning and every evening, a family ritual that bonded them even more tightly together. Having a pill that helps ease the burden of disease was something that every CF patient, every parent, had dreamed of when they first got their diagnosis of cystic fibrosis. Now Shelley got to do it for herself and her child. Personalized medicine had arrived on Shelley's doorstep, and just in time.

* * *

Other patients afflicted with cystic fibrosis have not been as lucky as Shelley. A few in our center passed away just prior to the availability of protein modulators, and more than a few over the decades prior. Others, like Sarah Murnaghan, have needed to undergo lung transplant and deal with the daily uncertainty that comes with chronic immune suppression. It is a horrible, deadly disease that robs children of their youth, and families of their offspring. But the spirit of investigation has transformed it from something unknown to something known, something untreatable to something treatable. And now, finally, to something on the cusp of being cured. Patients like Shelley, living longer, happier, and healthier, are becoming the norm, not the exception. The breath restored is indeed something magnificent.

Afterword

Progress in pulmonary medicine over the past few decades has been remarkable. In 1965 in the United States, 42 percent of the population were smokers, and today the rate of cigarette smoking is less than 14 percent and falling.[1] With declining smoking rates, a myriad of diseases should begin to abate, including lung cancer and COPD, heart disease, strokes, and many other types of cancers. This has, in fact, started to happen in earnest: in 2020, the biggest-ever single-year drop in cancer mortality was reported, 2.2 percent over 2016–2017, which was driven largely by an almost 5 percent decline in lung cancer.[2]

The medical advances in treating other lung diseases are no less notable. A CF patient in 1950 could expect to live a few years, while today life expectancy on average is forty-seven years.[3] The incidence of tuberculosis in America has never been lower. Cases of fatal asthma are way down, and finally there is a drug that has been shown to improve outcomes for those with idiopathic pulmonary fibrosis. The quality of our air in the United States has improved over the decades, despite a recent decline, with significant drops in all levels of pollutants.

Nonetheless, the threats to our breath are not taken seriously enough in many instances, and crises remain commonplace, from the recent devastating fires in California, the Amazon rainforest, and Australia, to vaping-related illnesses that have killed dozens, to the emergence of strange infections such as COVID-19. Only when we begin to understand all the many elements of the breath, and how they influence our well-being, can we address new crises in an effective fashion. The atmosphere of the Earth is perfectly fitted in its ability to sustain life, and vigilance to all aspects of our lung health is needed to ensure our future here on this planet, let alone on any others.

Fortunately, the future of pulmonary medicine looks bright, and at times resembles a science fiction novel. Personalized medicine has evolved to the point where lung cancer patients can undergo genetic analyses, which in turn can inform physicians of the best medicines to address their needs. With the promise of genetic manipulation, the day will come when a patient will walk into the clinic with cystic fibrosis and walk out a few hours later without any disease. The potential of stem cells, harvested from a person's own blood, will no doubt help us grow whole organs for transplant. A vaccine for tuberculosis could come any day, rendering the need for medications obsolete. Every pulmonary disease now has a horizon that was once unthinkable in its promise.

The health of our lungs in the absence of disease rests not in a pill or an injection, but in ensuring above all a healthy environment for our breath. Despite regression in this country at the federal level, many countries throughout the world are committed to lowering toxic greenhouse gases and lowering emissions from cars and power plants. Businesses are anticipating the future as well, with many auto compa-nies formulating plans to abandon gasoline-powered cars over the next few decades and opting for cleaner options such as electric and even hydrogen-powered cars. Individual states in the US are not waiting for leadership from Washington, DC, to emerge, but instead are formulat-ing their own plans for improving the environment. California wants 40 percent lower greenhouse gas emissions by 2030, compared to 1990 levels, and 80 percent lower by 2050.[4] People and governments at the local level understand the crisis of our atmosphere.

Worldwide, a healthy breath is going to ultimately depend on our being able to generate power without toxicity. Wind, solar, and geothermal sources of energy hold promise to reduce our reliance on fossil fuels. Fusion reactions can produce a great deal of energy with no carbon emissions and minimal toxic waste, and this technology has seen recent advances. A civilization freed from toxic fuel can be realized.

At a personal level, practicing common sense is always the best way to preserve healthy lung function. This includes avoiding tobacco smoke, ensuring your work and home environments have clean air, and cultivating an exercise, yoga, or other fitness plan. Individually, we must also actively support strong environmental protection and fight climate-change deniers. The stories in this book have shown us what science can achieve. The future is bright if we stick to the scientific principles of cause and effect and dedicated observation. For the health of our lungs and our bodies, and the future of our species and our planet, we must.

Acknowledgments

I would like to acknowledge my agent, Bonnie Solow, who was the first to see the potential in this book, provided immeasurable guidance, and never gave up on it. From here, Daryn Eller did an extraordinary job putting the pieces together in a fashion that worked. Ned Arnold offered numerous suggestions about ideas and organization. My editor, George Gibson, then provided insights that only somebody with immense experience could deliver. His patience and dedication to the project were beyond anything I could imagine. Thank you to all the scientists, doctors, and patients who took time from their busy days to speak with me. Their generosity never ceased to amaze me. I also need to acknowledge my family—my wife, Gudrun, daughter, Charlotte, and son, Julian—for their support. And finally to my mother, Johanna Pallotta Stephen, who helped me formulate the original idea for the project and has supported me in all my endeavors throughout my life.

Notes

Prologue: Lungs = Life

1. Holy Bible, Job 33:4 (New Revised Standard Version).

2. Ibid., John 20:22.

3. Ibid., Gen 2:7.

4. Julia Wolkoff, "Why Do So Many Egyptian Statues Have Broken Noses?" CNN.com, March 20, 2019, https://www.cnn.com/style/article/egyptian-statues-broken-noses-artsy/index.html.

5. Thich Nhat Hahn, *The Miracle of Mindfulness: An Introduction to the Practice of Meditation* (Boston, MA: Beacon Press, 1999), 15.

6. C. D. O'Malley, F. N. L. Poynter, and K. F. Russell, *William Harvey Lectures on the Whole of Anatomy, An Annotated Translation of Prelectiones Anatomiae Universalis* (Berkeley: University of California Press, 1961), 204.

7. Manoj K. Bhasin, Jeffrey A. Dusek, Bei-Hung Chang, et al., "Relaxation Response Induces Temporal Transcriptome Changes in Energy Metabolism, Insulin Secretion and Inflammatory Pathways," *PLOS One* 8, no. 5 (May 2013): e62817.

8. National Institutes of Health, "Cancer Stat Facts: Common Cancer Sites," National Cancer Institute, Surveillance, Epidemiology, and End Results Program website, accessed July 31, 2019, https://seer.cancer.gov/statfacts/html/common.html.

9. National Institutes of Health, "Estimates of Funding for Various Research, Condition, and Disease Categories," NIH website, https://report.nih.gov/categorical_spending.aspx.

10. David J. Lederer and Fernando J. Martinez, "Idiopathic Pulmonary Fibrosis," *New England Journal of Medicine* 378 (May 10, 2018): 1811–1823.

11. Rein M. G. J. Houben and Peter J. Dodd, "The Global Burden of Latent Tuberculosis Infection: A Re-Estimation Using Mathematical Modelling," *PLOS Medicine* 13 (October 25, 2016): e1002152.

12. Centers for Disease Control and Prevention, "Mortality Trends in the United States, 1900–2015," CDC website, accessed July 31, 2019, https://www.cdc.gov/nchs/data-visualization/mortality-trends/.

13. Romaine A. Pauwels and Klaus F. Rabe, "Burden and Clinical Features of Chronic Obstructive Pulmonary Disease (COPD)," *Lancet* 364, no. 9434 (August 2004): 613–620.

14. World Health Organization, "The Top 10 Causes of Death," WHO website, accessed May 8, 2020, https://www.who.int/news-room/fact-sheets/detail/the-top-10-causes -of-death.

15. Forum of International Respiratory Societies, *The Global Impact of Respiratory Disease*, 2nd ed. (Sheffield, UK: Sheffield, European Respiratory Society, 2017), 7.

16. World Health Organization, "Air Pollution," WHO website, accessed July 31, 2019, https://www.who.int/airpollution/en/.

Chapter 1: Oxygen, Then Existence

1. G. Brent Dalrymple, *Ancient Earth, Ancient Skies: The Age of Earth and Its Cosmic Surroundings* (Stanford, CA: Stanford University Press, 2004).

2. Bettina E. Schirrmeister, Muriel Gugger, and Philip C. J. Donoghue, "Cyanobacteria and the Great Oxidation Event: Evidence from Genes and Fossils," *Palaeontology* 58, no. 5 (September 2015): 769–785.

3. John Waterbury, in discussion with the author, July 2015.

4. John Waterbury, "Little Things Matter a Lot," *Oceanus Magazine*, March 11, 2005, https://www.whoi.edu/oceanus/feature/little-things-matter-a-lot/.

5. Christopher T. Reinhard, Noah J. Planavsky, Stephanie L. Olson, et al., "Earth's Oxygen Cycle and the Evolution of Animal Life," *PNAS* 113, no. 32 (August 9, 2016): 8933–8938.

6. Michael Melford, "Devonian Period," *National Geographic* website, accessed July 31, 2019, https://www.nationalgeographic.com/science/prehistoric-world/devonian/.

7. Keith S. Thomson, *Living Fossil: The Story of the Coelacanth* (New York: W. W. Norton, 1991), 19–49.

8. National Aeronautics and Space Administration, "Mars Oxygen In-Situ Resource Utilization Experiment (MOXIE)," NASA TechPort, accessed July 31, 2019, https:// techport.nasa.gov/view/33080.

9. National Aeronautics and Space Administration, "Planting an Ecosystem on Mars," NASA website, May 6, 2015, https://www.nasa.gov/feature/planting-an-ecosystem- on-mars.

Chapter 2: We Must Inhale and Exhale. But Why?

1. *Merriam-Webster* Online, s.v. "dum spiro, spero."

2. Roy Porter, *The Cambridge History of Medicine* (New York: Cambridge University Press, 2006), 78.

3. Daniel L. Gilbert, *Oxygen and Living Processes: An Interdisciplinary Approach* (New York: Springer-Verlag, 1981), 3.

4. Paula Findlen and Rebecca Bence, "A History of the Lungs," Stanford University website, Early Science Lab, https://web.stanford.edu/class/history13/earlysciencelab/body/lungspages/lung.html.

5. Andrew Cunningham, *The Anatomical Renaissance* (Abingdon, UK: Routledge, 2016), 61.

6. Saul Jarcho, "William Harvey Described by an Eyewitness (John Aubrey)," *American Journal of Cardiology* 2, no. 3 (September 1958): 381–384.

7. Thomas Wright, *William Harvey: A Life in Circulation* (Oxford, UK: Oxford University Press, 2013), xvii–xxi.

8. David G. Ashbaugh, D. Boyd Bigelow, Thomas L. Petty, et al., "Acute Respiratory Distress in Adults," *Lancet* 290, no. 7511 (August 12, 1967): 319–323.

9. Giacomo Bellani, John G. Laffey, Tai Pham, et al., "Epidemiology, Patterns of Care, and Mortality for Patients with Acute Respiratory Distress Syndrome in Intensive Care Units in 50 Countries," *JAMA* 315, no. 8 (2016): 788–800.

10. Roy G. Brower, Michael A. Matthay, Alan Morris, et al., "Ventilation with Lower Tidal Volumes as Compared with Traditional Tidal Volumes for Acute Lung Injury and the Acute Respiratory Distress Syndrome," *New England Journal of Medicine* 342 (May 4, 2000): 1301–1308.

11. Michael A. Mathay, Carolyn S. Calfee, Hanjing Zhuo, et al., "Treatment with Allogeneic Mesenchymal Stromal Cells for Moderate to Severe Acute Respiratory Distress Syndrome (START Study): A Randomised Phase 2a Safety Trial," *Lancet Respiratory Medicine* 7, no. 2 (February 2019): 154–162.

12. John B. West, "How Well Designed Is the Human Lung?" *American Journal of Respiratory and Critical Care Medicine* 173, no. 6 (2006): 583–584.

13. Adrian Bejan and Eden Mamut, *Thermodynamic Optimization of Complex Energy Systems* (Dordrecht, NL: Springer, 1999), 71.

Chapter 3: An Infant's Drive to Breathe

1. Mary Ellen Avery, MD, interview by Lawrence M. Gartner, American Academy of Pediatrics, Oral History Project, 2009. https://www.aap.org/en-us/about-the-aap/Gartner-Pediatric-History-Center/DocLib/Avery.pdf.

2. Amalie M. Kass and Eleanor G. Shore, "Mary Ellen Avery," *Harvard Magazine*, March-April 2018. https://harvardmagazine.com/2018/02/dr-mary-allen-avery.

3. John A. Clements and Mary Ellen Avery, "Lung Surfactant and Neonatal Respiratory Distress Syndrome," *American Journal of Respiratory and Critical Care Medicine* 157, no. 4 (1998): S59–S66.

4. John A. Clements, "Lung Surfactant: A Personal Perspective," *Annual Review of Physiology* 59 (1997): 1–21.

5. Clements, "Surface Tension of Lung Extracts," *Experimental Biology and Medicine* 95 (1957): 170–172.

6. Julius H. Comroe Jr., *Retrospectroscope: Insights into Medical Discovery* (Menlo Park CA: Von Gehr Press, 1977), 149–150.

7. Mary Ellen Avery and Jere Mead, "Surface Properties in Relation to Atelectasis and Hyaline Membrane Disease," *American Journal of Diseases of Children* 97 (May 1959): 517–523.

Chapter 4: The Extraordinary Healing Power of the Breath

1. Susan Scutti, "Drug Overdoses, Suicides Cause Drop in 2017 US Life Expectancy; CDC Director Calls It a 'Wakeup Call'," CNN Health (website), December 17, 2019, https://www.cnn.com/2018/11/29/health/life-expectancy-2017-cdc/index.html.

2. A. H. Weinberger, M. Gbedemah, A. M. Martinez, et al., "Trends in Depression Prevalence in the USA from 2005 to 2015: Widening Disparities in Vulnerable Groups," *Psychological Medicine* 48, no. 8 (June 2018): 1308–1315.

3. National Institutes of Health, "Major Depression," National Institute of Mental Health website, https://www.nimh.nih.gov/health/statistics/major-depression.shtml.

4. Donald Westerhausen, Anthony J. Perkins, Joshua Conley, et al., "Burden of Substance Abuse-Related Admissions to the Medical ICU," *Chest Journal* 157, no. 1 (January 2020), https://journal.chestnet.org/article/S0012-3692(19)33736-5/fulltext.

5. W. Andrew Baldwin, Brian A. Rosenfeld, Michael J. Breslow, et al., "Substance Abuse-Related Admissions to Adult Intensive Care," *Chest Journal* 103, no. 1 (January 1993), https://journal.chestnet.org/article/S0012-3692(16)38290-3/fulltext.

6. William J. Cromie, "Meditation Changes Temperatures," *Harvard Gazette*, April 18, 2002, https://news.harvard.edu/gazette/story/2002/04/meditation-changes-temperatures/.

7. "Fremont Kaiser Patient Told He's Dying Via Tele-Robot Doctor Visit," *CBSN Bay Area*, March 8, 2019, https://sanfrancisco.cbslocal.com/2019/03/08/kaiser-patient-told-dying-robot-doctor-video-call/.

8. BBC, "Religion," BBC website, https://www.bbc.co.uk/religion/religions/buddhism/.

9. Thich Nhat Hanh. *The Miracle of Mindfulness: An Introduction to the Practice of Meditation* (Boston: Beacon Press, 1999), 15.

10. Amy Weintraub, *Yoga for Depression: A Compassionate Guide to Relieve Suffering through Yoga* (New York: Broadway Books, 2004), 2.

11. Ibid., 3.

12. Jon Kabat-Zinn, *Meditation Is Not What You Think: Mindfulness and Why It's So Important* (New York: Hachette Books, 2018), 133.

13. Naykky Singh Ospina, Kari A. Phillips, Rene Rodriguez-Gutierrez, et al., "Eliciting the Patient's Agenda—Secondary Analysis of Recorded Clinical Encounters," *Journal of General Internal Medicine* 34 (2019): 36–40.

14. Abraham Verghese, Blake Charlton, Jerome P. Kassirer, et al., "Inadequacies of Physical Examination as a Cause of Medical Errors and Adverse Events: A Collection of Vignettes," *American Journal of Medicine* 128, no. 12 (December 2015): 1322–1324.

15. Robert L. Cowie, Diane P. Conley, Margot F. Underwood, and Patricia G. Reader, "A Randomised Controlled Trial of the Buteyko Technique as an Adjunct to Conventional Management of Asthma," *Respiratory Medicine* 102, no. 5 (May 2008): 726–732.

16. M. Thomas, R. K. McKinley, S. Mellor, et al., "Breathing Exercises for Asthma: A Randomised Controlled Trial," *Thorax* 64, no. 1 (2009): 55–61.

17. Nasrin Falsafi, "A Randomized Controlled Trial of Mindfulness Versus Yoga: Effects on Depression and/or Anxiety in College Students," *Journal of the American Psychiatric Nurses Association*, 22 (August 26, 2016): 483–497.

18. B. A. Van der Kolk, L. Stone, J. West, et al., "Yoga as an Adjunctive Treatment for Posttraumatic Stress Disorder: A Randomized Controlled Trial." *Journal of Clinical Psychiatry* 75 (2014): e559–565.

19. Arndt Büssing, Thomas Ostermann, Rainer Lüdtke, et al., "Effects of Yoga Interventions on Pain and Pain-Associated Disability: A Meta-Analysis. *Journal of Pain* 13, no. 1 (January 2012): 1–9.

20. Majoj K. Bhasin, Jeffrey A. Dusek, Bei-Hung Chang, et al., "Relaxation Response Induces Temporal Transcriptome Changes in Energy Metabolism, Insulin Secretion and Inflammatory Pathways," *PLOS One* 8, no. 5 (May 2013): e62817.

21. Nani Morgan, Michael R. Irwin, Mei Chung, et al., "The Effects of Mind-Body Therapies on the Immune System: Meta-Analysis," *PLOS One* 9, no. 7 (2014): e100903.

22. Wouter Van Marken Lichtenbelt, "Who Is the Iceman?" *Temperature* 4 (2017): 202–205.

23. Thich Nhat Hahn, *Stepping into Freedom: An Introduction to Buddhist Monastic Training* (Berkeley, CA: Parallax Press, 1997), 8.

24. Hahn, *Breathe, You Are Alive: The Sutra on the Full Awareness of Breathing* (Berkeley, CA: Parallax Press, 2008), i.

Chapter 5: A Window onto the Immune System

1. "Crawling Neutrophil Chasing Bacterium," YouTube video, 19:15, posted by Frantraf, May 20, 2006, https://www.youtube.com/watch?v=MgVPLNu_S-w.

2. Centers for Disease Control and Prevention, "Reported Cases and Deaths from Vaccine Preventable Diseases, United States, 1950–2013," CDC website, March, 2018, https://www.cdc.gov/vaccines/pubs/pinkbook/appendix/appdx-e.html.

3. Jean-François Bach, "The Effect of Infections on Susceptibility to Autoimmune and Allergic Diseases," *New England Journal of Medicine* 347 (September 19, 2002): 911–920.

4. Matthew F. Cusick, Jane E. Libbey, and Robert S. Fujinami, "Molecular Mimicry as a Mechanism of Autoimmune Disease," *Clinical Reviews in Allergy and Immunology* 42 (2012): 102–111.

5. Centers for Disease Control and Prevention. "Asthma as the Underlying Cause of Death," CDC website, https://www.cdc.gov/asthma/asthma_stats/asthma_underlying_death.html.

6. Javan Allison and Monique Cooper, *The Adventures of Javan and The 3 A's* (Amazon Digital Services, 2018).

7. Russell Noyes Jr., "Seneca on Death," *Journal of Religion and Health* 12 (1973): 223–240.

8. Marianna Karamanou and G. Androutsos, "Aretaeus of Cappadocia and the First Clinical Description of Asthma," *American Journal of Respiratory and Critical Care Medicine* 184 (2011): 1420–1421.

9. Mark Jackson, *Marcel Proust and the Global History of Asthma* (PowerPoint presentation), https://www.who.int/global_health_histories/seminars/presentation21.pdf.

10. Morrill Wyman, "Autumnal Catarrh," *The Boston Medical and Surgical Journal* 93 (1875): 209–212.

11. L. F. Haas, "Emil Adolph von Behring (1854-1917) and Shibasaburo Kitasato (1852-1931)," *Journal of Neurology, Neurosurgery & Psychiatry* 71, no. 1 (2001): 62.

12. Cormac Sheridan, "Convalescent Serum Lines Up as First-Choice Treatment for Coronavirus," *Nature Biotechnology News,* May 7, 2020, https://www.nature.com/articles/d41587-020-00011-1.

13. Arthur M. Silverstein, "Clemens Freiherr von Pirquet: Explaining immune complex disease in 1906," *Nature Immunology* 1 (2000): 453–455.

14. Maximilian A. Ramirez, "Horse Asthma Following Blood Transfusion," *JAMA* 73 (1919): 984–985.

15. Kimishige Ishizaka and Teruko Ishizaka, "Identification of IgE," *Journal of Allergy and Clinical Immunology* 137, no. 6 (June 2016): 1646–1650.

16. S. G. O. Johansson, "The Discovery of IgE," *Journal of Allergy and Clinical Immunology* 137, no. 6 (June 2016): 1671–1673.

17. Thomas A. E. Platts-Mills, Alexander J. Schuyler, Elizabeth A. Erwin, et al., "IgE in the Diagnosis and Treatment of Allergic Disease," *Journal of Allergy and Clinical Immunology* 137 (2016): 1662–1670.

18. Cristoforo Invorvaia, Marina Mauro, Marina Russello, et al., "Omalizumab, an Anti-Immunoglobulin E Antibody: State of the Art," *Drug Design, Development and Therapy* 8 (2014): 187–207.

19. Amelia Murray-Cooper, "Amount of Vegetation on Earth Increasing, BU-Led Study Shows," *Boston University Daily Free Press*, April 13, 2020, https://dailyfreepress.com/2019/03/22/amount-of-vegetation-on-earth-increasing-bu-led-study-shows/.

Chapter 6: The Lungs and the Common Good

1. "California Tuberculosis Patient Found, Arrested," *San Francisco Examiner*, July 29, 2014, https://www.sfexaminer.com/national-news/california-tuberculosis-patient -found-arrested/.

2. Associated Press, "Tuberculosis Patient Charged in Calif. for Not Taking Medication," *CBS News* Online, May 16, 2012, https://www.cbsnews.com/news/ tuberculosis-patient-charged-in-calif-for-not-taking-medication/.

3. S. M. Aciego, C. S. Riebe, and S. C. Hart, "Dust Outpaces Bedrock in Nutrient Supply to Montane Forest Ecosystems," *Nature Communications* 8 (2017): 14800.

4. National Aeronautics and Space Administration, "NASA Satellite Reveals How Much Saharan Dust Feeds Amazon's Plants," NASA website, February 22, 2015, https:// www.nasa.gov/content/goddard/nasa-satellite-reveals-how-much-saharan-dust -feeds-amazon-s-plants.

5. Nancy Tomes, *The Gospel of Germs: Men, Women, and the Microbe in American Life* (Cambridge, MA: Harvard University Press, 1998), 97.

6. William Firth Wells and Mildred Weeks Wells, "Air-Borne Infections," *JAMA* 107 (1936): 1698–1703.

7. Lydia Bourouiba, "Turbulent Gas Clouds and Respiratory Pathogen Emissions," *JAMA*, published online March 26, 2020, https://jamanetwork.com/journals/jama/ fullarticle/2763852.

8. Peter Disikes, "In the Cloud: How Coughs and Sneezes Float Farther Than You Think." *MIT News* Online, April 8, 2014, http://news.mit.edu/2014/coughs-and -sneezes-float-farther-you-think.

9. World Health Organization, "Modes of Transmission of Virus Causing COVID-19: Implications for IPC Precaution Recommendations," WHO website, accessed May 9, 2020, https://www.who.int/news-room/commentaries/ detail/modes-of-transmission-of-virus-causing-covid-19-implications-for-ipc -precaution-recommendations.

10. Sean Wei Xiang Ong, Yian Kim Tan, Po Ying Chia, et. al. "Air, Surface Environmental, and Personal Protective Equipment Contamination by Severe Acute Respiratory Syndrome Coronavirus 2 (SARS-CoV-2) From a Symptomatic Patient," *JAMA*, published online March 4, 2020, https://jamanetwork.com/journals/jama/ fullarticle/2762692.

11. Alice Yan, "Chinese expert who came down with Wuhan coronavirus after saying it was controllable thinks he was infected through his eyes," *South China Morning Post*, January 23, 2020, https://www.scmp.com/news/china/article/3047394/ chinese-expert-who-came-down-wuhan-coronavirus-after-saying-it-was.

12. Tom Paulson, "Epidemiology A Mortal Foe," *Nature* 502, no. 7470 (October 10, 2013): S2–S3.

13. World Health Organization, "Tuberculosis," WHO website, accessed September 18, 2018, https://www.who.int/news-room/fact-sheets/detail/tuberculosis.

14. I. Barberis, N. L. Bragazzi, L. Galluzzo, and M. Martini, "The History of Tuberculosis: From the First Historical Records to the Isolation of Koch's Bacillus," *Journal of Preventive Medicine and Hygiene* 58 (2017): E9–E12.

15. Anne C. Stone, Alicia K. Wilbur, Jane E. Buikstra, and Charlotte A. Roberts, "Tuberculosis and Leprosy in Perspective," *Yearbook of Physical Anthropology* 52 (2009): 66–94.

16. Kirsten I. Bos, Kelly M. Harkins, Alexander Herbig, et al., "Pre-Columbian Mycobacterial Genomes Reveal Seals as a Source of New World Human Tuberculosis," *Nature* 514 (2014): 494–497.

17. Clark Lawlor, *Consumption and Literature: The Making of the Romantic Disease* (Basingstoke, UK: Palgrave Macmillan, 2006), 111.

18. Arne Eggum, *Edvard Munch: Paintings, Sketches, and Studies* (New York: C. N. Potter, 1984), 46.

19. M. Monir Madkour, Kitab E. Al-Otaibi, and R. Al Swailem, "Historical Aspects of Tuberculosis" in *Tuberculosis* (Berlin Heidelberg: Springer-Verlag, 2004), 18.

20. Thomas M. Daniel, "Jean-Antoine Villemin and the Infectious Nature of Tuberculosis," *International Journal of Tuberculosis and Lung Disease* 19 (2015): 267–268.

21. Edward S. Golub, *The Limits of Medicine* (Chicago: University of Chicago Press, 1997), 93.

22. Alex Sakula, "Robert Koch: Centenary of the Discovery of the Tubercle Bacillus, 1882," *Canadian Veterinary Journal* 24, no. 4 (April 1983): 127–131.

23. Daniel M. Fox, "Social Policy and City Politics: Tuberculosis Reporting In New York, 1889–1900," *Bulletin of the History of Medicine* 49, no. 2 (Summer 1975): 169–195.

24. Godias J. Drolet and Anthony M. Lowell, *A Half Century's Progress Against Tuberculosis in New York City* (New York Tuberculosis and Health Association, 1952), https://www1.nyc.gov/assets/doh/downloads/pdf/tb/tb1900.pdf.

25. H. Sheridan Baketel and Arthur C. Jacobson, "Public Health," *The Medical Times*, 43 (June 1915): 200.

26. Corinne S. Merle, Katherine Fielding, Omou Bah Sow, et al., "A Four-Month Gatifloxacin-Containing Regimen for Treating Tuberculosis," *New England Journal of Medicine* 371 (October 23, 2014): 1588–1598.

27. Tasha Smith, Kerstin A. Wolff, and Liem Nguyen, "Molecular Biology of Drug Resistance in Mycobacterium Tuberculosis," *Current Topics in Microbiology and Immunology* 375 (2014): 53–80.

28. New York City Health Department, *New York City Bureau of Tuberculosis Control Annual Summary, 2018*, pdf file, https://www1.nyc.gov/assets/doh/downloads/pdf/tb/tb2018.pdf.

29. Karen Brudney and Jay Dobkin, "Resurgent Tuberculosis in New York City: Human Immunodeficiency Virus, Homelessness, and the Decline of Tuberculosis Control Programs," *The American Review of Respiratory Disease* 144, no. 4 (October 1991): 745–749.

30. Natalie Shure, "How New York Beat Its TB Epidemic," *The Daily Beast*, April 14, 2017, https://www.thedailybeast.com/how-new-york-beat-its-tb-epidemic.

31. New York City Health Department, *New York City Bureau of Tuberculosis Control Annual Summary, 2018*, pdf file, https://www1.nyc.gov/assets/doh/downloads/pdf/tb/tb2018.pdf.

32. Natalie Shure, "How New York Beat Its TB Epidemic," *The Daily Beast*, April 14, 2017, https://www.thedailybeast.com/how-new-york-beat-its-tb-epidemic.

33. World Health Organization, "Tuberculosis Country Profiles," WHO website, https://www.who.int/tb/country/data/profiles/en/.

34. World Health Organization, "Drug Resistant Tuberculosis," WHO website, https://www.who.int/tb/areas-of-work/drug-resistant-tb/en/.

35. "China Scientists Say SARS-Civet Cat Link Proved," *Science News*, January 20, 2007, https://www.reuters.com/article/us-china-sars/china-scientists-say-sars-civet-cat-link-proved-idUSPEK23793120061123.

36. "How Wildlife Trade is Linked to Coronavirus," YouTube video, 8:48, Vox, March 6, 2020, https://www.youtube.com/watch?v=TPpoJGYlW54.

37. David Cyranoski, "Mystery Deepens over Animal Source of Coronavirus," *Nature* Online, February 26, 2020, https://www.nature.com/articles/d41586-020-00548-w.

38. Sheri Fink and Mike Baker, "'It's Just Everywhere Already': How Delays in Testing Set Back the U.S. Coronavirus Response," *New York Times*, March 10, 2020, https://www.nytimes.com/2020/03/10/us/coronavirus-testing-delays.html.

39. Stephen Engelberg, Lisa Song, and Lydia DePillis, "How South Korea Scaled Coronavirus Testing While the U.S. Fell Dangerously Behind," *ProPublica*, March 15, 2020, https://www.propublica.org/article/how-south-korea-scaled-coronavirus-testing-while-the-us-fell-dangerously-behind.

Chapter 7: Nicotine Seduction and Stem Cells

1. Robert Evans, *A Brief History of Vice: How Bad Behavior Built Civilization* (New York: Plume, 2016), 152.

2. Laura Dwyer-Lindgren, Amelia Bertozzi-Villa, Rebecca W. Stubbs, et al., "Trends and Patterns of Differences in Chronic Respiratory Disease Mortality Among US Counties, 1980–2014," *JAMA* 318, no. 12 (September 26, 2017): 1136–1149.

3. Frederick Webb Hodge. *Handbook of American Indians North of Mexico Part 2* (Washington: United States Government Printing Office, 1910), 767.

4. Anthony Chute, *Tabaco* (London, 1595), https://archive.org/details/tabacco00chutgoog/page/n7/mode/2up?q=consumption.

5. Iain Milne, "A counterblaste to tobacco: King James's anti-smoking tract of 1616," *The Journal of the Royal College of Physicians of Edinburgh* 41 (2011): 89.

6. Sidney Andrews, *The South since the War: As Shown by Fourteen Weeks of Travel and Observation in Georgia and the Carolinas*, abr. ed. (Baton Rouge, LA: Louisiana State University Press, 2004), 87.

7. William Kremer, "James Buchanan Duke: Father of the Modern Cigarette," *BBC News Magazine*, November 13, 2012, https://www.bbc.com/news/magazine -20042217.

8. Rafael Laniado-Laborin, "Smoking and Chronic Obstructive Pulmonary Disease (COPD). Parallel Epidemics of the 21st Century," *International Journal of Environmental Research and Public Health* 6 (2009): 209–224.

9. Mariella De Biasi and John A. Dani, "Reward, Addiction, Withdrawal to Nicotine," *Annual Review of Neuroscience* 34 (2011): 105–130.

10. R. R. Baker, "Temperature Distribution Inside a Burning Cigarette, *Nature* 247 (1974): 405–406.

11. US Department of Health and Human Services, *A Report of the Surgeon General: How Tobacco Smoke Causes Disease: What It Means to You* (consumer booklet) (Atlanta, GA: US Department of Health and Human Services, Centers for Disease Control and Prevention, National Center for Chronic Disease Prevention and Health Promotion, Office on Smoking and Health, 2010), 30–44.

12. Stephen Babb, Ann Malarcher, Gillian Schauer, et al., "Quitting Smoking Among Adults—United States, 2000–2015," *Morbidity and Mortality Weekly Report* 65 (2017): 1457–1464.

13. G. R. Martin, "Isolation of a Pluripotent Cell Line from Early Mouse Embryos Cultured in Medium Conditioned by Teratocarcinoma Stem Cells," *Proceedings of the National Academy of Sciences of the United States of America* 78 (1981): 7634–7638.

14. M. J. Evans and M. H. Kaufman, "Establishment in Culture of Pluripotential Cells from Mouse Embryos, *Nature* 292 (1981): 154–156.

15. Kazutoshi Takahashi and Shinya Yamanaka, "Induction of Pluripotent Stem Cells from Mouse Embryonic and Adult Fibroblast Cultures by Defined Factors," *Cell* 126, no. 4 (August 25, 2006): 663–676.

16. Anjali Jacob, Michael Morley, Finn Hawkins, et al., "Differentiation of Human Pluripotent Stem Cells into Functional Lung Alveolar Epithelial Cells," *Cell Stem Cell* 21, no. 5 (October 5, 2017): 472–488.

17. Centers for Disease Control and Prevention, "Current Cigarette Smoking Among Adults in the United States," CDC website, November 18, 2019, https://www.cdc .gov/tobacco/data_statistics/fact_sheets/adult_data/cig_smoking/index.htm.

18. Teresa W. Wang, Andrea S. Gentzke, MeLisa R. Creamer, et al., "Tobacco Product Use and Associated Factors Among Middle and High School Students—United States, 2019," *Morbidity and Mortality Weekly Report* 68, no. 12 (December 6, 2019): 1–22, https://www.cdc.gov/mmwr/volumes/68/ss/ss6812a1.htm?s_cid=ss6812a1_w#T7 _down.

19. Hongying Dai and Adam M. Leventhal, "Prevalence of e-Cigarette Use Among Adults in the United States, 2014-2018," *JAMA* 322, no. 18 (2019): 1824–1827, https://jamanetwork.com/journals/jama/article-abstract/2751687.

20. Centers for Disease Control and Prevention, "Outbreak of Lung Injury Associated with Use of E-Cigarette, or Vaping, Products," CDC website, Smoking and

Tobacco Use, February 25, 2020, https://www.cdc.gov/tobacco/basic_information/e-cigarettes/severe-lung-disease.html#map-cases.

21. National Institutes of Health, "Nationwide Trends," National Institute on Drug Abuse website, June 2015, https://www.drugabuse.gov/publications/drugfacts/nationwide-trends.

22. Centers for Disease Control and Prevention, "Table 20: Use of Selected Substances in the Past Month Among Person Aged 12 Years and Over, by Age, Sex, Race, and Hispanic Origin, United States, Selected Years 2002–2017," pdf file, https://www.cdc.gov/nchs/data/hus/2018/020.pdf.

Chapter 8: Health Is Not the Absence of Disease: Climate Change

1. Steven R. James, R. W. Dennell, Allan S. Gilbert, et al., "Hominid Use of Fire in the Lower and Middle Pleistocene," *Current Anthropology* 30, no. 1 (February 1989): 1–26.

2. World Health Organization, "Air Pollution," WHO website, 2020, https://www.who.int/airpollution/en/.

3. Philip J. Landrigan, Richard Fuller, Nereus J. R. Acosta, et al., "The Lancet Commission on Pollution and Health," *Lancet* 391 (2018): 462–512.

4. Ibid., 465.

5. American Lung Association, "Particle Pollution," American Lung Association website, February 25, 2020, https://www.lung.org/our-initiatives/healthy-air/outdoor/air-pollution/particle-pollution.html.

6. Jim Morrison, "Air Pollution Goes Back Way Further Than You Think," *Smithsonian Magazine*, January 11, 2016, https://www.smithsonianmag.com/science-nature/air-pollution-goes-back-way-further-you-think-180957716/#BZ1IdR9y0MdRzJvy.99.

7. John Evelyn, *Fumigugium* (Exeter, UK: University of Exeter Press, 1976), https://archive.org/details/fumifugium00eveluoft/page/n5.

8. W. O. Henderson, *Industrial Britain Under the Regency* (Abingdon, UK: Routledge, 2006), 105.

9. Rob Baker, "'A Proper Pea-Souper'—The Dreadful London Smog of 1952," *Flashbak*, December 4, 2017, https://flashbak.com/proper-pea-souper-dreadful-london-smog-1952-391180/.

10. Edwin Kiester Jr., "A Darkness in Donora," *Smithsonian Magazine*, November 1999, https://www.smithsonianmag.com/history/a-darkness-in-donora-174128118/.

11. J. Lelieveld, J. S. Eans, M. Fnais, et al., "The Contribution of Outdoor Air Pollution Sources to Premature Mortality on a Global Scale," *Nature* 525 (2015): 367–371.

12. Deidre Lockwood, "California Farms Are a Silent but Sizable Source of Air Pollution," *Scientific American*, February 6, 2018, https://www.scientificamerican.com/article/california-farms-are-a-silent-but-sizable-source-of-air-pollution/.

13. State of Washington, Department of Ecology, *How Wood Smoke Harms Your Health*, pdf file, https://fortress.wa.gov/ecy/publications/publications/91br023.pdf.

14. "Emissions of Air Pollutants in the UK, 1970 to 2018—Particulate Matter" (PM10 and PM2.5), Department of Environment Food & Rural Affairs, gov.uk. https://www.gov.uk/government/publications/emissions-of-air-pollutants/emissions-of-air-pollutants-in-the-uk-1970-to-2018-particulate-matter-pm10-and-pm25.

15. American Lung Association, "State of the Air 2019," American Lung Association website, https://www.lung.org/assets/documents/healthy-air/state-of-the-air/sota-2019-full.pdf.

16. Diddier Prada, Jia Zhong, and Elena Colicino, "Association of Air Particulate Pollution with Bone Loss over Time and Bone Fracture Risk: Analysis of Data from Two Independent Studies," *Lancet Planetary Health* 1, no. 8 (2017): e337–e347.

17. Diana Younan, Andrew J. Petkus, Keith F. Widaman, et al., "Particulate Matter and Episodic Memory Decline Mediated by Early Neuroanatomic Biomarkers of Alzheimer's Disease," *Brain* 143, no. 1 (November 20, 2019): 289–302.

18. Hari Kumar and Kai Schultz, "Delhi, Blanketed in Toxic Haze, 'Has Become a Gas Chamber'," *New York Times*, November 7, 2017, https://www.nytimes.com/2017/11/07/world/asia/delhi-pollution-gas-chamber.html.

19. "Dangerous Air Pollution in India Forces Delhi Schools to Close for 2nd Time in Two Weeks," CBS News website, November 15, 2019, https://www.cbsnews.com/news/air-pollution-in-india-delhi-forces-schools-industry-closed-health-problems-today-2019-11-15/.

20. Landrigan et al., "The Lancet Com-mission on Pollution and Health," 462–512.

21. World Health Organization, "Air Pollution," WHO website, 2020, https://www.who.int/airpollution/en/.

22. Tony Kirby, "Heather Zar—Improving Lung Health for Children in Africa," *Lancet* 376 (September 4, 2010): 763.

23. Kirsten A. Donald, Michelle Hoogenhout, Christopher P. du Plooy, et al., "Drakenstein Child Health Study (DCHS): Investigating Determinants of Early Child Development and Cognition." *BMJ Paediatrics Open* 2, no. 1 (2018): e000282.

24. Ron Sender, Shai Fuchs, and Ron Milo, "Revised Estimates for the Number of Human and Bacteria Cells in the Body," *PLOS Biology* 14, no. 8 (2016): e1002533.

25. Miriam F. Moffatt and William O. C. M. Cookson, "The Lung Microbiome in Health and Disease," *Clinical Medicine* (London) 17, no. 6 (December 2017): 525–529.

26. Diane M. Gray, Lidija Turkovic, Lauren Willemse, et al., "Lung Function in African Infants in the Drakenstein Child Health Study. Impact of Lower Respiratory Tract Illness," *American Journal of Respiratory and Critical Care Medicine* 195, no. 2 (2017): 212–220.

27. W. James Gauderman, Robert Urman, Edward Avol, et al., "Association of Improved Air Quality with Lung Development in Children," *New England Journal of Medicine* 372, no. 10 (March 5, 2015): 905–913.

28. C. Arden Pope III, "Respiratory Disease Associated with Community Air Pollution and a Steel Mill, Utah Valley," *American Journal of Public Health* 79 (May 1989): 623–628.

29. C. Arden Pope III, Douglas L. Rodermund, and Matthew M. Gee, "Mortality Effects of a Copper Smelter Strike and Reduced Ambient Sulfate Particulate Matter Air Pollution," *Environmental Health Perspectives* 115, no. 5 (2007): 679–683.

30. United States Environmental Protection Agency, "International Treaties and Cooperation about the Protection of the Stratospheric Ozone Layer," USEPA website, September 24, 2018, https://www.epa.gov/ozone-layer-protection/international -treaties-and-cooperation-about-protection-stratospheric-ozone.

31. Jing Huang, Xiaochuan Pan, Xinbiao Guo, and Guoxing Li G, "Health Impact of China's Air Pollution Prevention and Control Action Plan: An Analysis of National Air Quality Monitoring and Mortality Data," *Lancet Planetary Health* 2, no. 7 (July 2018): e313–e323.

32. "Why Is India's Pollution Much Worse Than China's?" BBC News website, November 6, 2019, https://www.bbc.com/news/world-asia-50298972.

33. Steven Bernard and Amy Kazmin, "Dirty Air: How India Became the Most Polluted Country on Earth," *Financial Times*, December 11, 2018.

34. Ryan Wiser and Mark Bolinger, "2018 Wind Technologies Market Report," US Department of Energy, 8, https://eta-publications.lbl.gov/sites/default/files/wtmr_ final_for_posting_8-9-19.pdf.

35. California Energy Commission, "Renewable Energy," State of California website, 2020, https://www.energy.ca.gov/programs-and-topics/topics/renewable -energy.

36. Amanda Levin, "2017 Clean Energy by the Numbers: A State-by-State Look," National Resource Defense Council website, 2018, https://www.nrdc.org/experts/ amanda-levin/2017-clean-energy-by-the-numbers-a-state-by-state-look.

37. Tim Arango, Jose A. Del Real, and Ivan Penn, "5 Lessons We Learned From the California Wildfires," *New York Times*, November 4, 2019, https://www.nytimes .com/2019/11/04/us/fires-california.html.

Chapter 9: Exposures Unnecessary: Time Does Not Heal All Wounds

1. Hannah Holmes, *The Secret Life of Dust: From the Cosmos to the Kitchen Counter, the Big Consequences of Little Things* (Hoboken, NJ: Wiley, 2003), 8.

2. Anthony DePalma, *City of Dust: Illness, Arrogance, and 9/11* (Upper Saddle River, NJ: FT Press Science, 2010), 253.

3. Anthony DePalma, "Air Masks at Issue in Claims of 9/11 Illnesses," *New York Times*, June 5, 2006, https://www.nytimes.com/2006/06/05/nyregion/05masks.html.

4. Caroline Bankoff, "What We Know About How 9/11 Has Affected New Yorkers' Health, 15 Years Later," *New York Magazine*, September 10, 2016, http://nymag.com/

intelligencer/2016/09/15-years-later-how-has-9-11-affected-new-yorkers-health
.html.

5. DePalma, *City of Dust*, 30–31.

6. Adam Lisberg, "New Lung or WTC Cop Dies," *New York Daily News*, January 16, 2007, https://www.nydailynews.com/news/new-lung-wtc-dies-officer-stricken -months-ground-zero-article-1.263583.

7. Jonathan M. Samet, Allison S. Geyh, and Mark J. Utell, "The Legacy of World Trade Center Dust," *New England Journal of Medicine* 356, no. 22 (May 31, 2007): 2233–2236.

8. John Lehmann, "9/11 Ills Forcing Firemen off Job," *New York Post*, December 21, 2001, https://nypost.com/2001/12/21/911-ills-forcing-firemen-off-job/.

9. David J. Prezant, Michael Weiden, Gisela I. Banauch, et al., "Cough and Bronchial Responsiveness in Firefighters at the World Trade Center Site," *New England Journal of Medicine* 347, no. 11 (September 12, 2002): 806–815.

10. Hyun Kim, Robert Herbert, Philip Landrigan, et al., "Increased Rates of Asthma Among World Trade Center Disaster Responders," *American Journal of Industrial Medicine* 55, no. 1 (January 2012): 44–53.

11. Juan P. Wisnivesky, Susan L. Teitelbaum, Andrew C. Todd, et al., "Persistence of Multiple Illnesses in World Trade Center Rescue and Recovery Workers: A Cohort Study," *Lancet* 378, no. 9794 (September 3–9, 2011): 888–897.

12. Morton Lippmann, Mitchell D. Cohen, and Lung-Chi Chen, "Health Effects of World Trade Center (WTC) Dust: An Unprecedented Disaster with Inadequate Risk Management," *Critical Reviews in Toxicology* 45, no. 6 (2015): 492–530.

13. Samet et al., "The Legacy of World Trade Center Dust," 2233–2236.

14. Ankura Singh, Rachel Zeig-Owens, William Moir, et al., "Estimation of Future Cancer Burden Among Rescue and Recovery Workers Exposed to the World Trade Center Disaster," *JAMA Oncology* 4, no. 6 (2018): 828–831.

15. Centers for Disease Control and Prevention, "Program Statistics," World Trade Center Health Program website, February 7, 2020, https://www.cdc.gov/wtc/ ataglance.html.

16. Rosalie David, *The Manchester Mummy Project* (Manchester, UK: Manchester University Press, 1979), 97.

17. Irving J. Selikoff and Douglas H. K. Lee, *Asbestos and Disease* (London: Academic Press, 1978), 4.

18. United Nations of Roma Victrix, "Asbestos in the Roman Empire," UNRV.com, https://www.unrv.com/economy/asbestos.php.

19. Irving J. Selikoff and Morris Greenberg, "A Landmark Case in Asbestosis," *JAMA* 265, no. 7 (1991): 898–901.

20. W. E. Cooke, "Fibrosis of the Lungs Due to the Inhalation of Asbestos Dust," *British Medical Journal* 147, no. 2 (1924): 147.

21. Peter Bartrip, *The Way from Dusty Death* (London: The Athlone Press, 2001), 12.

22. Miriam Haritz, *An Inconvenient Deliberation* (Alphen aan den Rijn, NL: Kluwer Law International, 2011), 78.

23. Alex Strauss, "Mesothelioma Takes Life of Merlin Olsen," Surviving Mesothelioma website, March 12, 2010, https://survivingmesothelioma.com/mesothelioma-takes-life-of-merlin-olsen/.

24. Jasek M. Mazurek, Girija Syamlal, John M. Wood, et al., "Malignant Mesothelioma Mortality—United States, 1999–2015," *Morbidity and Mortality Weekly Report* 66, no. 8 (March 3, 2017): 214–218.

25. Tim Povtak, "US Geological Survey: 750 Metric Tons of Asbestos Imported in 2018," Mesothelioma Center, asbestos.com website, https://www.asbestos.com/news/2019/03/26/asbestos-imports-2018-chloralkali/.

26. Andrew E. Kramer, "City in Russia Unable to Kick Asbestos Habit," *New York Times*, July 13, 2013, https://www.nytimes.com/2013/07/14/business/global/city-in-russia-unable-to-kick-asbestos-habit.html.

27. Tim Povtak, "Asbestos Mining in Russia Still Fuels the Economy in Some Cities," Mesothelioma Center, asbestos.com website, https://www.asbestos.com/news/2013/07/16/asbestos-mining-russia-fuels-economy/.

Chapter 10: Curing the Incurable

1. Daniel T. Montoro, Ada, L. Haber, Moshe Biton, et al., "A Revised Airway Epithelial Hierarchy Includes CFTR-Expressing Ionocytes," *Nature* 560 (2018): 319–324.

2. David J. Lederer and Fernando J. Martinez, "Idiopathic Pulmonary Fibrosis," *New England Journal of Medicine* 378 (2018): 1811–1823.

3. Paul J. Wolters, Timothy S. Blackwell, Oliver Eickelberg, et al., "Time for a Change: Is Idiopathic Pulmonary Fibrosis Still Idiopathic and Only Fibrotic?" *Lancet Respiratory Medicine* 6, no. 2 (2018): 154–160.

4. Harold R. Collard, Jay H. Ryu, William W. Douglas, et al., "Combined Corticosteroid and Cyclophosphamide Therapy Does Not Alter Survival in Idiopathic Pulmonary Fibrosis," *Chest* 125, no. 6 (June 2004): 2169–2174.

5. Paul Noble, "Idiopathic Pulmonary Fibrosis. Proceedings of the 1st Annual Pittsburgh International Lung Conference, October 2002," *American Journal of Respiratory Cell and Molecular Biology* 29 (2003): S1–105.

6. Ganesh Raghu, Kevin K. Brown, Williamson Z. Bradford, et al., "A Placebo-Controlled Trial of Interferon Gamma-1b in Patients with Idiopathic Pulmonary Fibrosis," *New England Journal of Medicine* 350 (2004): 125–133.

7. Shreekrishna M. Gadekar, US Patent # US3974281A, 5-Methyl-1-phenyl-2-(1H)-pyridone compositions and methods of use, Google Patents, https://patents.google.com/patent/US3974281A/en?assignee=AFFILIATED+MED+RES.

8. Solomon B. Margolin, US Patent # US5310562, Composition and Method for Reparation and Prevention of Fibrotic Lesions, pdf file, https://patentimages.storage.googleapis.com/6e/5b/23/d9202c3ecdef2d/US5310562.pdf.

9. S. N. Iyer, J. S. Wil, M. Schiedt, et al., "Dietary Intake of Pirfenidone Ameliorates Bleomycin-Induced Lung Fibrosis in Hamsters," *Journal of Laboratory and Clinical Medicine* 125, no. 6 (May 31, 1995): 779–785.

10. H. Taniguchi, M. Ebina, T. Kondoh, et al., "Pirfenidone in Idiopathic Pulmonary Fibrosis," *European Respiratory Journal* 35 (2010): 821–829.

11. Paul W. Noble, Carlo Albera, Williamson Z. Bradford, et al., "Pirfenidone in Patients with Idiopathic Pulmonary Fibrosis (CAPACITY): Two Randomised Trials," *Lancet* 377, no. 9779 (May 21–27, 2011): 1760–1769.

12. Talmadge E. King Jr., Williamson Z. Bradford, Socorro Castro-Bernardini, et al., "ASCEND Study Group. A Phase 3 Trial of Pirfenidone in Patients with Idiopathic Pulmonary Fibrosis," *New England Journal of Medicine* 370 (May 29, 2014): 2083–2092.

13. Luca Richeldi, Roland M. du Bois, Ganesh Raghu, et al., "Efficacy and Safety of Nintedanib in Idiopathic Pulmonary Fibrosis," *New England Journal of Medicine* 370 (May 29, 2014): 2071–2082.

14. Dianhua Jiang, Jiurong Liang, Juan Fan, et al., "Regulation of Lung Injury and Repair by Toll-Like Receptors and Hyaluronan," *Nature Medicine* 11 (2005): 1173–1179.

15. Jiurong Liang, Yanli Zhang, Ting Xie, et al., "Hyaluronan and TLR4 Promote Surfactant Protein C-Positive Alveolar Progenitor Cell Renewal and Prevent Severe Pulmonary Fibrosis in Mice," *Nature Medicine* 22 (2016): 1285–1293.

16. Wendy Henderson, "How a Runner in His 70s Cheats Pulmonary Fibrosis," *Pulmonary Fibrosis News*, March 24, 2017, https://pulmonaryfibrosisnews.com/2017/03/24/78-year-old-runner-shows-how-he-cheats-pulmonary-fibrosis/.

17. Joan E. Nichols, Jean A. Niles, Stephanie P. Vega, and Joaquin Cortiella, "Novel *in vitro* Respiratory Models to Study Lung Development, Physiology, Pathology and Toxicology," *Stem Cell Research and Therapy* 4 (2013): S7.

Chapter 11: Getting Personal with the Lungs

1. Alan Blum, "Alton Ochsner, MD, 1896–1981: Anti-Smoking Pioneer," *Ochsner Journal* 1 (1999): 102–105.

2. Luca Paoletti, Bianca Jardin, Matthew Carpenter, et al., "Current Status of Tobacco Policy and Control," *Journal of Thoracic Imaging* 27 (2012): 213–219.

3. Alton Ochsner and Michael DeBakey, "Primary Pulmonary Malignancy: Treatment by Total Pneumonectomy; Analysis of 79 Collected Cases and Presentation of 7 Personal Cases," *Surgery, Gynecology and Obstetrics* 1, no. 3 (1939): 435–445.

4. Richard Doll and A. Bradford Hill, "Smoking and Carcinoma of the Lung," *British Medical Journal* 2 (1950): 739–748.

5. S. S. Birring and M. D. Peake, "Symptoms and the Early Diagnosis of Lung Cancer," *Thorax* 60 (2005): 268–269.

6. American Lung Association, "Lung Cancer Fact Sheet," American Lung Association website, https://www.lung.org/lung-health-and-diseases/lung-disease-lookup/lung-cancer/resource-library/lung-cancer-fact-sheet.html.

7. National Institutes of Health, "Cancer Stat Facts: Common Cancer Sites," National Cancer Institute, Surveillance, Epidemiology, and End Results website, accessed July 31, 2019, https://seer.cancer.gov/statfacts/html/common.html.

8. National Institutes of Health, "Estimates of Funding for Various Research, Condition, and Disease Categories," Research Portfolio Online Reporting Tools website, accessed July 31, 2019, https://report.nih.gov/categorical_spending.aspx.

9. Christopher A. Haiman, Daniel O. Stram, Lynn R. Wilkens, et al., "Ethnic and Racial Differences in the Smoking-Related Risk of Lung Cancer," *New England Journal of Medicine* 354 (January 26, 2006): 333–342.

10. American Lung Association, "Tobacco Use in Racial and Ethnic Populations," ALA website, https://www.lung.org/stop-smoking/smoking-facts/tobacco-use-racial-and-ethnic.html.

11. Centers for Disease Control and Prevention, "Current Cigarette Smoking Among Adults in the United States," CDC website, https://www.cdc.gov/tobacco/data_statistics/fact_sheets/adult_data/cig_smoking/index.htm.

12. National Cancer Institute, "National Cancer Act of 1971," National Cancer Institute website, https://dtp.cancer.gov/timeline/flash/milestones/M4_Nixon.htm.

13. Leena Gandhi, Delvys Rodriguez-Abreu, Shirish Gadgeel, et al., "Pembrolizumab plus Chemotherapy in Metastatic Non–Small-Cell Lung Cancer," *New England Journal of Medicine* 378 (May 31, 2018): 2078–2092.

Chapter 12: The Breath and the Voice

1. James Stewart, "Singing: The First Art," *VPR Classical*, March 9, 2020, https://www.npr.org/podcasts/491502270/timeline.

2. John D. Clough, *To Act as a Unit: The Story of the Cleveland Clinic*, 4th ed. (Cleveland Clinic Press, 2005), 11.

3. Ibid., 12.

4. M. H. Mellish, *Collected Papers of the Mayo Clinic, volume XIII* (Philadelphia, PA: W.B. Saunders Company, 1922), 1275.

5. Teddi Barron, "After Being Mute, Iowa State University Graduating Senior Speaks with a New Voice," Iowa State University News Service, December 13, 2011, https://www.news.iastate.edu/news/2011/dec/KevinNeff.

6. "Once Gasping for Breath, Now Breathing Easy," Cleveland Clinic Foundation *Health Extra*, January 2004, https://my.clevelandclinic.org/ccf/media/files/Head_Neck/head_neck_testimonial.pdf.

7. Peter Densen, "Challenges and Opportunities Facing Medical Education," *Transactions of the American Clinical and Climatological Association* 122 (2011): 48–58.

8. Wim Lucassen, Geert-Jan Geersing, Petra M. G. Erkens, et al., "Clinical Decision Rules for Excluding Pulmonary Embolism: A Meta-Analysis," *Annals of Internal Medicine* 155 (2011): 448–460.

9. Casey Ross and Ike Swetlitz, "IBM's Watson Supercomputer Recommended 'Unsafe and Incorrect' Cancer Treatments, Internal Documents Show," *STAT News*, July 25, 2018.

10. Christopher D. Hanks, Jonathan Parson, Cathy Benninger, et al., "Etiology of Dyspnea in Elite and Recreational Athletes," *Physician and Sportsmedicine* 40, no. 2 (2012): 28–33.

11. Nalin J. Patel, Carol Jorgensen, and Joan Kuhn, "Concurrent Laryngeal Abnormalities in Patients with Paradoxical Vocal Fold Dysfunction," *Otolaryngology—Head Neck Surgery* 130 (2004): 686–689.

Chapter 13: The Miracle of Lung Transplant

1. United States Renal Data System, *2019 USRDS Annual Data Report: Epidemiology of Kidney Disease in the United States* (Bethesda, MD: National Institutes of Health, National Institute of Diabetes and Digestive and Kidney Diseases, 2019).

2. Ashok Jain, Jorge Reyes, Randeep Kashyap, et al., "Long-Term Survival After Liver Transplantation in 4,000 Consecutive Patients at a Single Center," *Annals of Surgery* 323, no. 4 (October 2000): 490–500.

3. US Department of Health and Human Services, "Organ Procurement and Transplantation Network National Data," Organ Procurement and Transplantation Network website, accessed January 16, 2020, https://optn.transplant.hrsa.gov/data/view-data-reports/national-data/#.

4. International Society for Heart and Lung Transplantation, "International Thoracic Organ Transplant (TTX) Registry Data Slides," International Society for Heart and Lung Transplantation website, https://ishltregistries.org/registries/slides.asp.

5. Federico Venuta and Dirk van Raemdonck, "History of Lung Transplantation," *Journal of Thoracic Disease* 9, no. 12 (December 2017): 5458–5471.

6. Tom Meek, "This Month in 1980: 33 Years Since Cyclosporine Demonstrated its Potential as an Immunosuppressant," *PMLiVE*, March 25, 2013, http://www.pmlive.com/pharma_news/33_years_since_cyclosporine_demonstrated_its_potential_as_an_immunosuppressant_468977.

7. J. M. Bill Nelems, Anthony S. Rebuck, Joel D. Cooper, et al., "Human Lung Transplantation," *Chest* 78 (1980): 569–573.

8. "#45—World's First Successful Single Lung Transplant," YouTube video, 15:48, posted by UHN Toronto, November 4, 2013, https://www.youtube.com/watch?v=UIVtrdKlPWg.

9. US Department of Health and Human Services, "Organ Procurement and Transplantation Network National Data," Organ Procurement and Transplantation

Net-work website, accessed January 16, 2020, https://optn.transplant.hrsa.gov/data/view-data-reports/national-data/#.

10. "#45—World's First Successful Single Lung Transplant," YouTube video, 15:48, posted by UHN Toronto, November 4, 2013, https://www.youtube.com/watch?v=UIVtrdKlPWg.

11. Robin Vox, Wim A. Wuyts, Olivier Gheysens, et al., "Pirfenidone in Restrictive Allograft Syndrome After Lung Transplantation: A Case Series," *American Journal of Transplantation* 18, no. 12 (December 2018): 3045–3059.

12. B. Smeritschnig, P. Jaksch, A. Kocher, et al., "Quality of Life After Lung Transplantation: A Cross-Sectional Study," *Journal of Heart and Lung Transplantation* 24, no. 4 (April 2005): 474–480.

13. US Department of Health and Human Services, "Organ Procurement and Transplantation Network National Data for Lung Donors Recovered 1988–2017," Organ Procurement and Transplantation Network website, accessed January 16, 2020, https://optn.transplant.hrsa.gov/data/view-data-reports/national-data/#.

Chapter 14: The Greatest Medical Story Never Told

1. Bruce C. Marshall, "Survival Trending Upward But What Does This Really Mean?" Cystic Fibrosis Foundation, *CF Community Blog*, November 16, 2017, https://www.cff.org/CF-Community-Blog/Posts/2017/Survival-Trending-Upward-but-What-Does-This-Really-Mean/.

2. James Littlewood, "The History of Cystic Fibrosis," Cystic Fibrosis Medicine website, www.cfmedicine.com.

3. Stephanie Clague, "Dorothy Hansine Andersen," *Lancet Respiratory Medicine* 2, no. 3 (March 1, 2014): 184–185.

4. Dorothy H. Andersen, "Cystic Fibrosis of the Pancreas and Its Relation to Celiac Disease," *American Journal of Diseases of Children* 56, no. 2 (1938): 344–399.

5. Walter F. Naedele, "Dr. Milton Graub, 90, Pediatrician," *Philadelphia Inquirer*, July 19, 2010, https://www.inquirer.com/philly/obituaries/20100719_Dr__Milton_Graub__90__pediatrician.html.

6. L. C. Tsui, M. Buchwald, D. Barker, et al., "Cystic Fibrosis Locus Defined by a Genetically Linked Polymorphic DNA Marker," *Science* 230 (1985): 1054–1057.

7. J. M. Rommens, M. C. Ianuzzi, B. Kerem, et al., "Identification of the Cystic Fibrosis Gene: Chromosome Walking and Jumping," *Science* 245 (1989): 1059–1065.

8. "Warrren Alpert Foundation Prize Symposium," YouTube video, 4:00:20, posted by Harvard Medical School, October 5, 2017, https://www.youtube.com/watch?v=rVE8yB_RA9k.

9. P. M. Quinton. "Chloride Impermeability in Cystic Fibrosis," *Nature* 301, no. 5899 (February 3, 1983): 421–2.

10. Carl Zimmer, "Ancient Viruses Are Buried in Your DNA," *New York Times*, October 4, 2017, https://www.nytimes.com/2017/10/04/science/ancient-viruses-dna-genome.html.

11. "Warrren Alpert Foundation Prize Symposium," YouTube video, 4:00:20, posted by Harvard Medical School, October 5, 2017, https://www.youtube.com/watch?v=rVE8yB_RA9k.

12. Robert F. Higgins, Sophie LaMontagne, and Brent Kazan, "Vertex Pharmaceuticals and the Cystic Fibrosis Foundation: Venture Philanthropy Funding for Biotech," Harvard Business School Case 808-005, October 2007 (revised July 2013).

13. Bonnie W. Ramsey, Jane Davies, N. Gerard McElvaney, et al., "A CFTR Potentiator in Patients with Cystic Fibrosis and the G551D Mutation," *New England Journal of Medicine* 365, no. 18 (November 3, 2011): 1663–1672.

14. Claire E. Wainwright, J. Stuart Elborn, Bonnie W. Ramsey, et al., "Lumacaftor-Ivacaftor in Patients with Cystic Fibrosis Homozygous for Phe508del CFTR," *New England Journal of Medicine* 373 (2015): 220–231.

15. Jennifer L. Taylor-Cousar, Anne Munck, Edward F. McKone, et al., "Tezacaftor–Ivacaftor in Patients with Cystic Fibrosis Homozygous for Phe508del," *New England Journal of Medicine* 377, no. 21 (November 23, 2017): 2013–2023.

16. Peter G. Middleton, Marcus A. Mall, Pavel Drevinek, et al., "Elexacaftor-Tezacaftor-Ivacaftor for Cystic Fibrosis with a Single Phe508del Allele," *New England Journal of Medicine* 381, no. 19 (November 7, 2019): 1809–1819.

17. Clinic encounters with the author, December 18, 2019, and January 8, 2020.

Chapter 15: Cystic Fibrosis, the Most Heartbreaking Lung Disease

1. Janet Murnaghan, *Saving Sarah: One Mother's Battle Against the Health Care System to Save Her Daughter's Life* (New York: St. Martin's Press, 2018).

2. https://www.foxnews.com/us/case-of-dying-10-year-old-prompts-federal-call-for-review-of-child-organ-transplant-rules.

3. US Department of Health and Human Services, "Organ Procurement and Transplantation Network, National Data," Organ Procurement and Transplantation Network website, https://optn.transplant.hrsa.gov/data/view-data-reports/national-data/#.

4. Thomas M. Egan and Leah B. Edwards, "Effect of the Lung Allocation Score on Lung Transplantation in the United States," *Journal of Heart and Lung Transplantations* 35, no. 4 (April 2016): 433–439.

5. Karen Ladin and Douglas W. Hanto, "Rationing Lung Transplants—Procedural Fairness in Allocation and Appeals," *New England Journal of Medicine* 369, no. 7 (August 15, 2013): 599–601.

6. Janet Murnaghan, *Saving Sarah: One Mother's Battle Against the Health Care System to Save Her Daughter's Life* (New York: St. Martin's Press, 2018).

7. Chris Welch and Zain Asher, "With Just Weeks Left, Sarah Fights the System for Life-Saving Pair of Lungs," CNN Online, May 27, 2013, https://www.cnn.com/2013/05/27/health/pennsylvania-girl-lungs/index.html.

8. Ibid.

9. Brett Norman and Jason Millman, "Sebelius Ordered to Make Exception on Transplant," *Politico*, June 5, 2013, https://www.politico.com/story/2013/06/sarah-murnaghan-lung-transplant-ruling-kathleen-sebelius-092299.

10. Howard Panitch, e-mail message to the author with transcript of speech, October 3, 2014.

11. Sarah Murnaghan, "Acceptance Speech for Shining Star Award," (meeting of the Cystic Fibrosis Foundation, Philadelphia, PA, February 2014).

12. J. deSante, A. Caplan, B. Hippen, et al., "Was Sarah Murnaghan Treated Justly?" *Pediatrics* 134, no. 1 (July 2014): 155–162.

13. Bonnie W. Ramsey, Margaret S. Pepe, Joanne M. Quan, et al., "Intermittent Administration of Inhaled Tobramycin in Patients with Cystic Fibrosis," *New England Journal of Medicine* 340 (January 7, 1999): 23–30.

14. Henry J. Fuchs, Drucy S. Borowitz, David H. Christiansen, et al., "Effect of Aerosolized Recombinant Human DNase on Exacerbations of Respiratory Symptoms and on Pulmonary Function in Patients with Cystic Fibrosis," *New England Journal of Medicine* 331 (September 8, 1994): 637–642.

Afterword

1. Centers for Disease Control and Prevention, "Current Cigarette Smoking Among Adults in the United States," CDC website, https://www.cdc.gov/tobacco/data_statistics/fact_sheets/adult_data/cig_smoking/index.htm.

2. Stacy Simon, "Facts & Figures 2020 Reports Largest One-year Drop in Cancer Mortality," American Cancer Society, January 8, 2020.

3. Bruce C. Marshall, "Survival Trending Upward But What Does This Really Mean?" Cystic Fibrosis Foundation, *CF Community Blog*, November 16, 2017, https://www.cff.org/CF-Community-Blog/Posts/2017/Survival-Trending-Upward-but-What-Does-This-Really-Mean/.

4. State of California, "Climate Change Programs," California Air Resources Board website, https://www.arb.ca.gov/cc/cc.htm.

Image Credits

Index

Note: Page numbers in **bold** indicate figures.